University

LECTURE

Series

Volume 3

Why the Boundary of a Round Drop Becomes a Curve of Order Four

A. N. Varchenko
P. I. Etingof

American Mathematical Society
Providence, Rhode Island

1991 *Mathematics Subject Classification.* Primary 30C99, 31A25, 58F07, 76S05.

Library of Congress Cataloging-in-Publication Data

Varchenko, A. N. (Aleksandr Nikolaevich)

[Pochemu granitsa krugloĭ kapli prevrashchaetsia v inversnyĭ obraz ellipsa? English]

Why the boundary of a round drop becomes a curve of order four/A. N. Varchenko and P. I. Etingof.

p. cm. —(University lecture series, ISSN 1047-3998; 3)

Includes bibliographical references.

ISBN 0-8218-7002-5

1. Boundary value problems. 2. Curves, Elliptic. 3. Fluid dynamics–Mathematical models. I. Etingof, P. I. (Pavel I.), 1969– . II. Title. III. Series: University lecture series (Providence, R.I.).

QA379.V36 1992 92-20985

515′.35—dc20 CIP

Printed in the United States of America

The paper used in this book is acid-free and falls within the guidelines
established to ensure permanence and durability. ∞

This publication was typeset using AMS-TEX,
the American Mathematical Society's TEX macro system.

10 9 8 7 6 5 4 3 2 1 97 96 95 94 93 92

Contents

Preface vii

1. Mathematical model 1

1.1. Filtration flow of an incompressible fluid 1

1.2. The moving boundary problem 2

2. First integrals of boundary motion 5

2.1. Richardson's integrability theorem 5

2.2. Reconstruction of a domain from the values of its moments, and the inverse problem of two-dimensional potential theory 6

2.3. Results on the uniqueness of a domain with given moments (potential) 7

2.4. The result of injection does not depend on the order of work of the sources and sinks 9

Problems 10

3. Algebraic solutions 11

3.1. Algebraic and abelian domains 11

3.2. Algebraic solutions 11

3.3. The Cauchy transform and its properties 12

3.4. Singularity correspondence theorem 12

3.5. Proof of the theorem on algebraic solutions 14

3.6. Construction of algebraic solutions 14

3.7. Examples 16

Problems 18

4. Contraction of a gas bubble 21

4.1. Formulation of the problem 21

4.2. The inclusion property 22

4.3. Contraction of a convex domain 23

4.4. Contraction points 24

4.5. An analogue of Richardson's theorem 25

4.6. Dynamics of the gravity potential 27

4.7. The gravity potential as the solution of a boundary value problem . 27

4.8. Proof of the main theorem 28

4.9. Self-similar solutions 29

4.10. Asymptotics of contraction 30

4.11. Several sources 31

Problems 32

5. Evolution of a multiply connected domain 35

5.1. Statement of the problem 35

5.2. Integrals of motion 35

5.3. Algebraic solutions 36

5.4. Riemann's theorem 37

5.5. Singularity correspondence 38

5.6. Proof of the theorem on algebraic solutions 39

5.7. Construction of solutions 39

5.8. Reconstruction of an annular domain from its moments 40

Problems 42

6. Evolution with topological transformations 43

6.1. Weak solutions of the evolution problem 43

6.2. The simplest algebraic solutions 46

6.3. Weak solutions of the contraction problem 50

6.4. A sufficient condition of a breakup of a symmetric domain 51

Problems 52

7. Contraction problem on surfaces 55

7.1. Physical motivation 55

7.2. Potential and contraction points 55

7.3. Calculation of the generalized gravity potential on the surface 56

7.4. Breakup of the boundary on symmetric surfaces of revolution 58

Problems 60

Answers and clues to the problems 61

A few open questions 69

References 71

Preface

In the forties P. Ya. Polubarinova-Kochina and P. P. Kufarev studied the problem of evolution of a round oil spot surrounded by water when oil is extracted from a well inside the spot (Figure 1). It turned out that the boundary of the spot remains an algebraic curve of degree four in the course of evolution. This curve is the image of an ellipse under a reflection with respect to a circle. In 1950 Kufarev managed to generalize this property: if initially the oil spot is the image of the unit disk under a conformal map given by a rational function of the complex coordinate in the disk, then it retains this property in the course of evolution.

In 1972 S. Richardson found an infinite series of first integrals of motion of the spot. He proved that the integral of any harmonic function over the oil domain changes linearly in time. This allowed Richardson to give a new proof of the invariance of rationality and an effective method to construct explicit solutions.

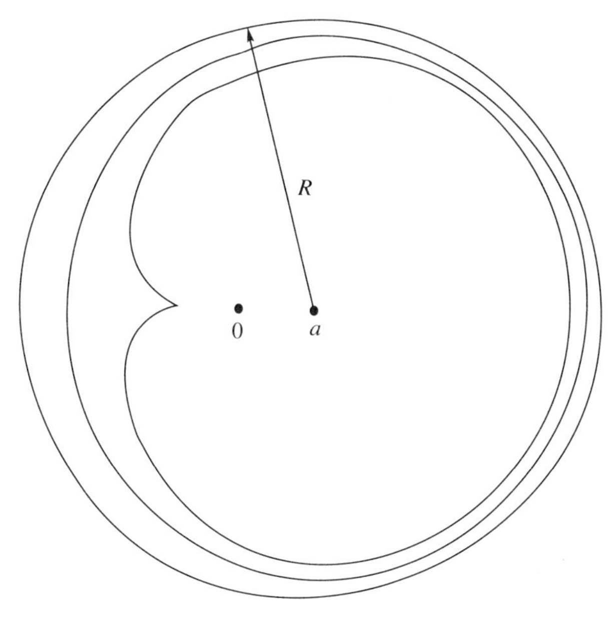

FIGURE 1

It was realized recently that this approach can be extended to multiply connected domains. In this case, the unit circle is replaced by a half of a complex algebraic curve with a specified M-structure, and the conformal map is given by a path integral of a meromorphic differential on the curve that has no zeros or poles on the chosen half.

Below we discuss these and other interesting mathematical subjects that arose recently in the theory of fluid flows with a moving boundary.

This text is an extended version of the first author's talk at a Moscow Mathematical Society meeting about the results of the second author. The authors gratefully acknowledge the crucial influence of Prof. V. M. Entov who introduced them to the circle of physical problems under consideration. The authors would also like to thank Prof. V. I. Arnold, D. Ya. Kleinbock, Prof. I. M. Krichever, and Dr. A. I. Shnirelman for very useful discussions, and Dr. Y. Peres for reading the English translation of the manuscript and making important remarks.

1. Mathematical Model

1.1. Filtration flow of an incompressible fluid. Consider a planar flow of a homogeneous fluid through a homogeneous porous medium. Such a flow is modeled by a time-dependent vector field in the plane, $\mathbf{v} = (v_1, v_2)$, which is called the fluid velocity.

Suppose that the fluid is incompressible. This property is expressed by the differential equation

$$\operatorname{div} \mathbf{v} = \frac{\partial v_1}{\partial x} + \frac{\partial v_2}{\partial y} = 0. \tag{1.1}$$

The main law of filtration is Darcy's law.[1] It states that the fluid velocity is proportional to the pressure gradient:

$$\mathbf{v} = -\kappa \operatorname{grad} p. \tag{1.2}$$

Here p is the pressure, $\kappa > 0$ is a proportionality coefficient. It is known that κ is inverse proportional to the dynamical fluid viscosity μ: $\kappa = k/\mu$. The coefficient k depends solely on the properties of the porous medium.

Thus, the fluid velocity is a potential vector field:

$$\mathbf{v} = \operatorname{grad} \Phi, \qquad \Phi = -\kappa p, \tag{1.3}$$

and its potential Φ is a harmonic function:

$$\Delta \Phi = \frac{\partial^2 \Phi}{\partial x^2} + \frac{\partial^2 \Phi}{\partial y^2} = 0. \tag{1.4}$$

Introduce the complex coordinate $z = x + iy$.

Let the region of flow contain sources and sinks. Let their coordinates and rates be $z_1, \dots, z_n$ and $q_1, \dots, q_n$ respectively. This means that near the point z_j

$$\mathbf{v}(z) = \frac{q_j}{2\pi(\bar{z} - \bar{z}_j)} + \text{smooth vector-function}, \tag{1.5}$$

or, equivalently,

$$\Phi(z) = \frac{q_j}{2\pi} \log|z - z_j| + \text{smooth function}. \tag{1.6}$$

---

[1]This law was discovered experimentally in 1856 by A. Darcy, a French engineer, when designing a system of public fountains.

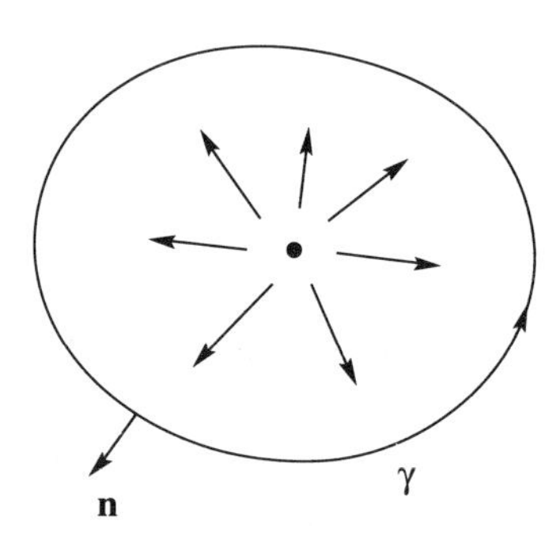

FIGURE 2

For any simple contour γ that goes around z_j counterclockwise,

$$\int_\gamma (\mathbf{v}, \mathbf{n})\, dl = q_j, \tag{1.7}$$

(see Figure 2). Geometrically, this means that the area occupied by the amount of fluid that passes through the contour in a unit time interval equals q_j.

1.2. The moving boundary problem. We want to study the motion of the boundary between two fluids saturating a porous medium. Here is the mathematical model we will refer to.

Consider a region of one fluid that is surrounded by another fluid of zero viscosity. It is assumed that the pressure is constant in the nonviscous fluid and continuous across the boundary between the fluids. Therefore, the potential is constant throughout each connected component of the boundary.

For now suppose the viscous fluid region to be simply connected (without holes) and bounded. Then we can assume that the potential vanishes on the boundary, since adding a constant to the potential does not change the velocity field:

$$\Phi|_{\partial D} = 0. \tag{1.8}$$

This assumption together with conditions (1.4) and (1.6) defines uniquely the velocity potential Φ_D for any domain D that contains the points $z_1, \ldots, z_n$. It is simply a linear combination of Green's functions of D with poles at these points. Its existence is guaranteed by the standard theorems about the Dirichlet problem; see **[1]**.

The law of motion of the boundary is dictated by Darcy's law:
Every point $z \in \partial D$ *moves with velocity* $\operatorname{grad} \Phi_D(z)$.

MAIN PROBLEM. *Describe the evolution of the viscous fluid region, given its initial shape and the positions and rates of sources and sinks.*

REMARK. Similar problems arise when describing the evolution of the boundary between fluid and gas, polymer material and air, crystallic and molten substance, and so on.

In practice the viscosities of oil, gas, and water are in the following approximate proportion: oil : water = 1 to 10, water : gas = 10 to 100. Therefore, the assumption about zero viscosity is most realistic for gas.

The problem we have stated can be solved approximately by a version of Euler's method for ordinary differential equations as follows. Calculate the velocity potential $\Phi_D(0)$ of the initial domain $D(0)$ and translate each point $\gamma \in D(0)$ by the vector $\Delta t \cdot \operatorname{grad} \Phi(\gamma)$, Δt being a small time interval. We will obtain the domain $D(\Delta t)$. Repeating this process, we will define domains $D(2\Delta t)$, $D(3\Delta t)$, ..., $D(n\Delta t)$, and this sequence will be an approximation to the desired family of domains $D(t)$.

2. First Integrals of Boundary Motion

2.1. Richardson's integrability theorem. The problem of the boundary evolution has a remarkable infinite series of first integrals. This property was discovered by Richardson in 1972.

THEOREM [2]. *Let $D(t)$ be the viscous fluid region at a time t, and let u be an arbitrary harmonic function in the plane. Then*

$$\frac{d}{dt}\int_{D(t)} u\,dx\,dy = \sum_{j=1}^{n} q_j u(z_j). \tag{2.1}$$

EXAMPLE. The area of $D(t)$ changes in time at the rate $\sum q_j$ $(u = 1)$.

Let us call the integral $\int_D u\,dx\,dy$ the *moment* of the domain D with respect to the harmonic function u. This theorem describes the dynamics of the moments of $D(t)$ in time. In particular, if the rates q_j are constant, all the moments change linearly in time.

PROOF OF THE THEOREM. It is seen from Figure 3 that

$$\int_{D(t+\Delta t)} u\,dx\,dy = \int_{D(t)} u\,dx\,dy + \int_{\partial D(t)} v_b \Delta t\,dl + o(\Delta t), \quad \Delta t \to 0,$$

where $v_b = (\operatorname{grad}\Phi, \mathbf{n}) = \pm|\operatorname{grad}\Phi|$ is the scalar value of the boundary velocity. Therefore

$$\frac{d}{dt}\int_{D(t)} u\,dx\,dy = \int_{\partial D(t)} u(\operatorname{grad}\Phi, \mathbf{n})\,dl.$$

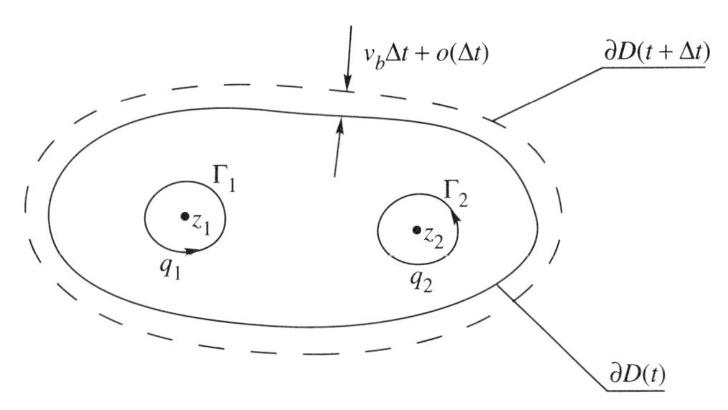

FIGURE 3

Adding to the right-hand side the term $-\int_{\partial D(t)} \Phi(\operatorname{grad} u, \mathbf{n})\, dl$, which equals zero because the potential vanishes at the boundary, we obtain, using Green's formula,

$$\begin{aligned}\frac{d}{dt}\int_{D(t)} u\,dx\,dy &= \int_{\partial D(t)} u(\operatorname{grad}\Phi, \mathbf{n})\,dl - \int_{\partial D(t)} \Phi(\operatorname{grad} u, \mathbf{n})\,dl \\ &= \int_E \operatorname{div}(u\operatorname{grad}\Phi - \Phi\operatorname{grad} u)dx\,dy \\ &\quad + \sum_{j=1}^{n}\int_{\Gamma_j} (u\operatorname{grad}\Phi - \operatorname{grad} u, \mathbf{n})dl,\end{aligned}$$

where E is the domain $D(t)$ with n disks of a radius ε with centers at z_j cut out. Integral over E equals zero because the functions u and Φ are harmonic. Taking into account the asymptotics of the potential at z_j, we get

$$\int_{\Gamma_j} (\Phi\operatorname{grad} u, \mathbf{n})\,dl \to 0, \qquad \int_{\Gamma_j} (u\operatorname{grad}\Phi, \mathbf{n})\,dl \to u(z_j)q_j$$

as $\varepsilon \to 0$, which implies the statement of the theorem.

REMARK. In the theorem it is sufficient to require that the function u be harmonic in some domain that contains $D(t)$ together with its boundary.

Examples of harmonic functions: $1, x, y, xy, x^2 - y^2$.

Since harmonic functions on the plane are real (imaginary) parts of holomorphic functions, for every holomorphic function $f(z)$ we have

$$\frac{d}{dt}\int_{D(t)} f(z)dx\,dy = \sum_{j=1}^{n} q_j f(z_j).$$

In particular, for any integer $k \geq 0$

$$\frac{d}{dt}\int_{D(t)} z^k dx\,dy = \sum_{j=1}^{n} q_j z_j^k.$$

So, the moments of a domain with respect to all harmonic functions can be expressed in terms of Richardson's moments $M_k(D) = \int_D z^k dx\,dy$, since any holomorphic function expands into a Taylor series.

2.2. Reconstruction of a domain from the values of its moments, and the inverse problem of two-dimensional potential theory. Richardson's theorem suggests the following method of solving the evolution problem. Since the initial domain is known, we can evaluate its moments. The moments of the domain $D(t)$ at any time t can then be found from (2.1):

$$\int_{D(t)} u\,dx\,dy = \int_{D(0)} u\,dx\,dy + \sum_{j=1}^{n} u(z_j)\int_0^t q_j(\tau)d\tau. \tag{2.2}$$

So, it is sufficient to be able to reconstruct the domain from its moments. It gives rise to the following general problem.

RICHARDSON'S MOMENT PROBLEM. *Reconstruct a simply connected bounded domain from the sequence of its moments.*

This problem has a nice physical interpretation.

INVERSE PROBLEM OF TWO-DIMENSIONAL POTENTIAL THEORY. *Reconstruct the shape of an infinite homogeneous cylinder from the gravity field that it creates outside itself.*

Let D be the orthogonal cross-section of the cylinder, and let w be the complex coordinate in the section plane.

It is easy to calculate that the gravity force equals the gradient of the logarithmic gravity potential

$$\Pi_D(w) = \frac{1}{2\pi}\int_D \log|z-w|\,dx\,dy, \tag{2.3}$$

up to a factor that depends solely on the density of the material. To know this function outside the domain (or just in a neighborhood of infinity) means the same as to know the moments of the domain, since

$$\Pi_D(w) = \frac{M_0}{2\pi}\log|w| + \sum_{k=1}^{\infty}\frac{1}{2\pi k}\mathrm{Re}\left(\frac{M_k}{w^k}\right), \tag{2.4}$$

M_k being the moment of D with respect to the function z^k.

2.3. Results on the uniqueness of a domain with given moments (potential). The inverse problem of two-dimensional potential theory has been studied intensively because of its applications to geophysics. Here is one of the earliest mathematical results about it.

DEFINITION. A domain D is called *starlike* with respect to a point P if it contains the chord PQ together with every point $Q \in D$.

THEOREM (P. S. Novikov, 1938, [3]). *If two bounded domains are starlike with respect to a common point, and their outer gravity potentials are the same, then these domains coincide.*

Still, there exist distinct simply connected bounded domains with equal moments. Let us give a construction of such domains. Consider two arbitrary simply connected bounded domains D_1 and D_2 whose boundaries intersect transversally at some point P (Figure 4, p. 8). Choose a point Q in the intersection of the domains. Denote by $D_1(t)$ and $D_2(t)$ the domains that are produced from D_1 and D_2 by injection of fluid from a source at Q at some rate (the same for both domains). Consider annular domains $E_1(t) = D_1(t)\backslash D_1$, $E_2(t) = D_2(t)\backslash D_2$. Their moments are equal, according to formula (2.2). Excluding from both $E_1(t)$ and $E_2(t)$ the curvilinear tetragon Δ, we obtain simply connected domains G_1 and G_2 with equal moments.

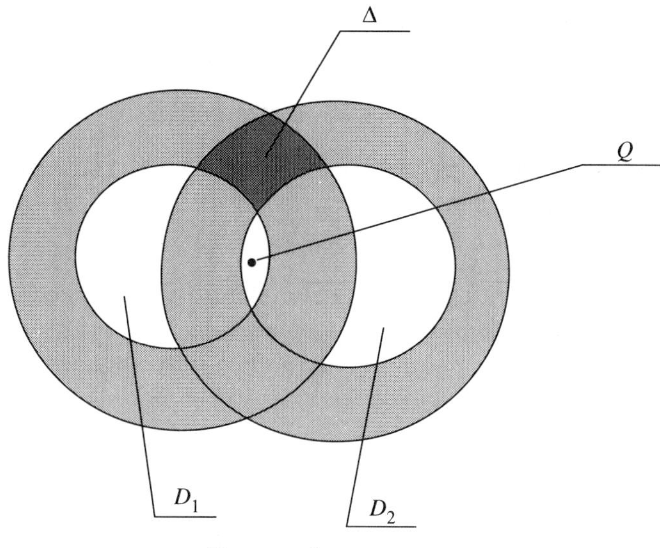

FIGURE 4

REMARKS. 1. An example of two domains with equal moments was first published by Sakai in 1978 [**4**]. The authors have been told that a similar example was earlier constructed by P. S. Novikov. A. M. Gabrielov constructed an example of two simply connected polygons with equal moments [**28**].

2. One can show that for any N there exists a set of N domains with equal moments. Moreover, one can construct an infinite set of domains with this property (see Problem 3 and [**22**]).

3. The constructed domains with equal moments have singularities on the boundary. This defect can be corrected. Let us inject equal amounts of fluid into both domains from sources at the vertices of Δ. The two domains obtained in this way will still have equal potentials (by Richardson's theorem), and their boundaries will be analytic curves.

THEOREM [**5**] (on local uniqueness of a domain with prescribed moments). *Let $D(s)$ be a smooth family of simply connected domains with equal moments. Then* $D(s) = \text{const}$.

PROOF. $\frac{d}{ds}\int_{D(s)} u\,dx\,dy = \int_{\partial D(s)} uv\,dl$ (Figure 3), v being the speed of deformation of the boundary (a function on the boundary). Suppose that v is not identically zero. Choose a harmonic function u so that $|u - v| < \varepsilon$ on $\partial D(s)$. For this purpose, let us take the domain D^* of points whose distance to $D(s)$ is less than δ (a small number that depends on ε) and then solve the Dirichlet problem in D^*: $\Delta u = 0$ in D^*, $u|_{\partial D^*} = v^*$, where v^* is a function that is close to v. The solution will be a harmonic function in a neighborhood of $D(s)$ that is close to v on the boundary of $D(s)$ in the supremum metric. When ε is small enough, $\int_{\partial D(s)} vu\,dl > 0$. But this integral must equal zero by the invariance of the moments. This is a contradiction. Hence, v is identically zero, i.e. the family is constant.

This theorem has a corollary that guarantees that any nice family of domains with the right moment properties has to be the right one, i.e. the solution of the injection problem.

COROLLARY. *Let $D(t)$, $t \in [0, T)$, be a smooth family of simply connected domains whose moments change according to* (2.1) *as t increases. Then $D(t)$ is a solution of the injection problem with the initial domain $D(0)$.*

PROOF. Assume this is false, and $D(t)$ is not a solution. We can construct the solution of the injection problem with the initial domain $D(0)$ and sources at points z_j of rates q_j. This will give us another family of domains $D'(t)$ with moments changing according to (2.1). For $0 \le s \le t$, denote by $D(s, t)$ the result of injection into the domain $D(s)$ during the time $t - s$ from the system of sources described above. By (2.1), the moments of $D(s, t)$ do not depend on s. Since $D(s, t)$ is a smooth family of domains, by the local uniqueness theorem it must be constant. Then $D(0, t) = D(t, t)$. But $D(0, t) = D'(t)$, $D(t, t) = D(t)$. So, we have proved that $D(t) = D'(t)$, i.e. $D(t)$ is a solution.

EXAMPLE **[6]**. Let D be a bounded simply connected domain, A be a positive number, and P be a point on the plane. Assume that for any harmonic function u, $\int_D u\,dx\,dy = Au(P)$. Then D is the circle of area A with center at P. To prove this it is enough to observe that the domains D_φ obtained by rotation of D by an angle φ, as well as D, satisfy the above condition. Therefore, according to the latter theorem, $D_\varphi = D$ for all φ.

2.4. The result of injection does not depend on the order of work of the sources and sinks. Define a strategy of extraction as a vector-function of rates of sources:

$$\mathbf{q}(t) = (q_1(t), q_2(t), \dots, q_n(t)),$$

and the *total* of this strategy for a time T as the vector

$$\mathbf{Q} = \left(\int_0^T q_1(t)dt, \int_0^T q_2(t)dt, \dots, \int_0^T q_n(t)dt\right).$$

Suppose we have a smooth family of strategies on a time interval $[0, T]$: $\mathbf{q}_s(t)$. Assume that their total is independent of s. Consider the evolution of any initial domain produced by the strategies $\mathbf{q}_s(t)$. At the time T we will obtain some final fluid domains $D_s(T)$.

COROLLARY (of the local uniqueness theorem). *The final domain does not depend on the strategy* (*on* s).

In particular, the schedule of work of the sources and sinks does not matter: the transformations of an initial domain produced by different sources are commutative.

Problems. 1. Find the center of mass of the domain obtained from a disk by extraction through an eccentrically situated sink during a prescribed time interval.

2. Prove: if two bounded convex domains have equal outer gravity potential, they are identical.

3. Give an example of any finite number of distinct simply connected bounded domains with equal moments.

3. Algebraic Solutions

3.1. Algebraic and abelian domains. Evolution of domains has a remarkable property: algebraic domains transform to algebraic ones.

According to the theorem of B.Riemann, any simply connected domain on the plane whose boundary contains more than one point is conformally equivalent to the disk. More precisely, let $K = \{\zeta \in \mathbf{C} : |\zeta| < 1\}$ be the unit disk. There exists a complex analytic function f defined on K that realizes a one-to-one correspondence between K and D. Such a function is not unique, since the disk has nontrivial conformal automorphisms. The group of conformal maps of the disk onto itself consists of the three-parameter family of functions $z = \frac{\zeta - a}{1 - \bar{a}\zeta} e^{i\theta}$, where $a \in \mathbf{C}$, $|a| < 1$, $\theta \in [0, 2\pi)$. In order to specify a conformal map from K to D, it is sufficient to fix the image of the origin in D and the angle $\arg f'(0)$.

Let us call any conformal map of the disk onto a domain a uniformization map of this domain.

The class of domains whose uniformization map is a rational function, $f(\zeta) = \frac{P(\zeta)}{Q(\zeta)}$, where P, Q are polynomials, is of a special interest. Let us call such a domain algebraic, and let us define the degree of an algebraic domain as the number $\deg f = \max(\deg P, \deg Q)$, provided P and Q are relatively prime. An algebraic domain D is bounded by an algebraic curve of degree $2 \deg D$, but not every domain with an algebraic boundary is algebraic itself. Algebraic domains of a bounded degree are determined by a finite number of parameters — the coefficients of the polynomials.

It also makes sense to consider abelian domains, i.e those whose uniformization map has rational derivative $f'(\zeta)$. Let us call $\deg f'$ the multiplicity of an abelian domain.

3.2. Algebraic solutions. Let $D(t)$ be the evolving domain.

Theorem (S. Richardson, P. P. Kufarev, [2], [12]). 1. *If $D(0)$ is an algebraic domain of degree d then $D(t)$ is an algebraic domain of degree no higher than $d + n$.*

2. *If $D(0)$ is an abelian domain of multiplicity k then $D(t)$ is an abelian domain of multiplicity no higher than $k + 2n$.*

So, there exist solutions of the evolution problem such that $D(t)$ is algebraic (abelian) for all t. Finding such solutions is a finite-dimensional problem. Solutions having this property are dense in the space of all solutions in any reasonable topology (for example, C^k, C^∞, analytic), because any bounded simply connected domain can be approximated by algebraic domains as closely as desired.

3.3. The Cauchy transform and its properties. The main role in the proof and applications of the above theorem belongs to the *Cauchy transform* of a domain D:

$$h_D(w) = \frac{1}{\pi}\int_D \frac{dxdy}{w-z}, \qquad w \notin D. \tag{3.1}$$

$h_D(w)$ is the moment of D with respect to the harmonic function $1/\pi(w-z)$, therefore, by Richardson's theorem we have

$$\frac{d}{dt}h_{D(t)}(w) = \sum_{j=1}^{n} \frac{q_j}{\pi(w-z_j)}. \tag{3.2}$$

The Cauchy transform has the following properties.

1. h_D is an analytic function outside D with zero limit at infinity.

2. $h_D(w) = \frac{1}{\pi}\sum_{k=0}^{\infty} \frac{M_k(D)}{w^{k+1}}$.

3. Let $F_D(w)$ be the gravity field of the cylinder with section D (see §2). Then $F_D(w) = \overline{h_D(w)}$. This follows from formulas (2.3) and (2.4).

4. $h_D(w) = \frac{1}{2\pi i}\int_{\partial D} \frac{\bar{z}dz}{w-z}$. This formula is derived from (3.1) and Green's formula.

5. $\frac{1}{2\pi i}\int_{\partial D} \frac{\bar{z}-h_D(z)}{w-z} = 0$. This equality follows from property 4 and the Cauchy theorem.

6. $h_{D(t)}(w) = h_{D(0)}(w) + \sum_{j=1}^{n} \frac{q_j t}{\pi(w-z_j)}$ for constant rates q_j. If q_j depend on t then $q_j t$ must be replaced by $\int_0^t q_j(\tau)\,d\tau$. This property can be obtained by integrating (3.2).

7. $h_{D(t)}$ is rational if and only if so is $h_{D(0)}$. $\frac{dh_{D(t)}}{dw}$ is rational if and only if so is $\frac{dh_{D(0)}}{dw}$. These are obvious consequences of property 6.

As a corollary of property 1, let us note that the closures of any set of domains with equal moments have a nontrivial intersection. Indeed, these domains must have the same Cauchy transform. This function has to be analytic outside the closure of each of them. If the intersection were empty, this function would be analytic everywhere, i.e. would have to be identically zero, which is impossible.

REMARK. Open domains with equal moments can have empty intersection (see the solution of problem 3 of §2).

3.4. Singularity correspondence theorem.

THEOREM (Richardson, Gustafsson [**6**], [**7**]). *Let f be a uniformization map of a domain D that maps the unit disk onto D. Then*

1. *The function* $\psi(\zeta) = \overline{f(1/\bar{\zeta})} - h_D(f(\zeta))$ *continues to a holomorphic function in the unit disk.*

2. *The function* $f(\zeta)$ *is rational if and only if the function* $h_D(w)$ *is rational.*

3. *The function* $df(\zeta)/d\zeta$ *is rational if and only if the function* $dh_D(w)/dw$ *is rational.*

4. *If* D *is an algebraic domain then the functions* f *and* h_D *have the same degree. If* D *is an abelian domain then the functions* $df/d\zeta$ *and* dh_D/dw *have the same degree.*

PROOF. 1. The main tool of the proof is the Sokhotskiĭ-Plemelj formula (see **[8]**).

Sokhotskiĭ-Plemelj formula. Let D be a bounded domain on the plane, let ϕ be a continuous function on ∂D, and

$$F^+(z) = \int_{\partial D} \frac{\phi(\xi)}{\xi - z} d\xi, \qquad z \in D;$$

$$F^-(z) = \int_{\partial D} \frac{\phi(\xi)}{\xi - z} d\xi, \qquad z \notin D.$$

Then for any point z_0 on the boundary

$$\lim_{z \to z_0} F^+(z) - \lim_{z \to z_0} F^-(z) = 2\pi i \phi(z_0).$$

In particular, if $F^-(z)$ is identically zero, the function $\phi(\zeta)$ extends to a holomorphic function in D (namely, to the function $F^+(z)/2\pi i$).

According to the Sokhotskiĭ-Plemelj formula and property 5 of the Cauchy transform, the function $\bar{z} - h_D(z)$ extends holomorphically inside D. Hence, the function $\overline{f(\zeta)} - h_D(f(\zeta))$, defined on the unit circle, analytically continues inside the unit disk. The analytic continuation of $\overline{f(\zeta)}$ inside the disk is the analytic function with singularities $\overline{f(1/\bar{\zeta})}$, because on the unit circle $\zeta = 1/\bar{\zeta}$. This completes the proof of 1.

2. If the function f is rational then the function $\overline{f(1/\bar{\zeta})}$ is meromorphic inside the unit disk. This implies that $h_D(f(\zeta))$ extends to a meromorphic function in the disk (according to statement 1 of the theorem). So, h_D extends inside D with a finite number of poles. Since h_D is regular outside of D, it has to be a rational function.

Conversely, if h_D is a rational function, the function $h_D(f(\zeta))$ is meromorphic inside the unit disk, hence so is $\overline{f(1/\bar{\zeta})}$. This implies that the function $f(\zeta)$ is meromorphic outside the unit disk. Since by definition this function is holomorphic inside the disk, it has to be rational.

3. The proof is analogous to that of 2.

4. According to statement 1 of the theorem, there is the following one-to-one correspondence between singular points of the continuation of the function f to the exterior of the unit circle and the continuation of the

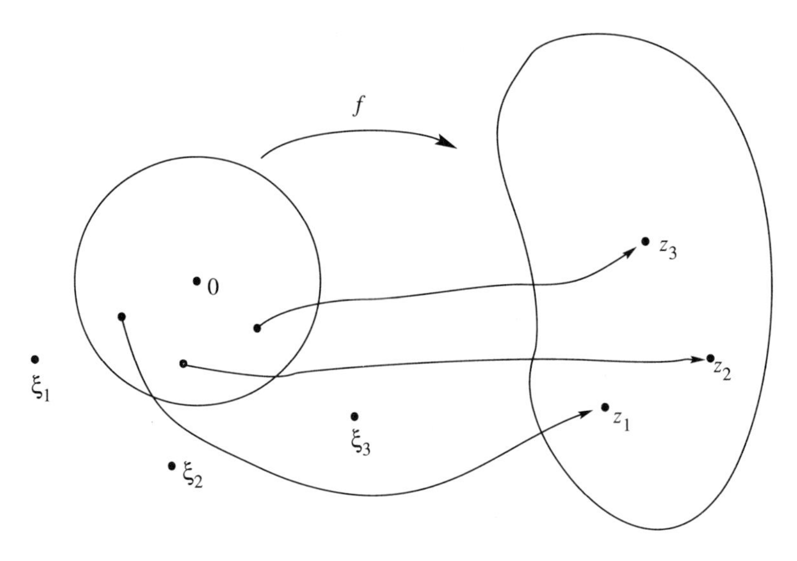

FIGURE 5

function h_D inside the domain D (Figure 5). Let ξ be a singularity of f, and let $\eta = \bar{\xi}^{-1}$ be its image under reflection in the unit circle. Then $f(\eta)$ is a singularity of h_D of the same kind (this explains the title of this section and the name of the theorem). So, the functions f and h_D, $df/d\zeta$ and dh_D/dz have equal number of poles of every order. This implies that their degrees are also equal, because the degree of a rational function equals the sum of orders of its poles.

3.5. Proof of the theorem on algebraic solutions. According to properties 6 and 7 of the Cauchy transform, the functions h_D and dh_D/dw are rational for all t if and only if they are rational at $t = 0$, and $\deg h_{D(t)} - \deg h_{D(0)} \leq n$, $\deg \frac{dh_{D(t)}}{dw} - \deg \frac{dh_{D(0)}}{dw} \leq 2n$, where $\deg h$ stands for the degree of a rational function h. By the singularity corrrespondence theorem, this implies that the uniformization map f_t of the domain $D(t)$ and its derivative $df_t/d\zeta$ are rational if and only if they are rational at $t = 0$, and $\deg f_t - \deg f_0 \leq n$, $\deg \frac{df_t}{dw} - \deg \frac{df_0}{d\zeta} \leq 2n$. This completes the proof of the theorem.

3.6. Construction of algebraic solutions. The singularity correspondence theorem reduces the problem of constructing solutions in case of an algebraic (abelian) initial domain to solving a finite system of nonlinear equations that express the coefficients of the uniformization map in terms of the known coefficients of the Cauchy transform. These equations are deduced from the condition that the principal parts of the functions $\overline{f(1/\bar{\zeta})}$ and $h_D(f(\zeta))$ at their singular points inside the unit disk coincide. Consider first the simplest case when there is a unique singular point.

THEOREM (L. A. Galin, [9], 1945). *Let the only source be situated at the origin, and let* $f_t(0) = 0$ *for all* t. *Then, if* $f_0(\zeta)$ *is a polynomial of degree* d, *so is* $f_t(\zeta)$ *for all* t.

PROOF. The only singularity of $f_0(\zeta)$ is at infinity, and it is a pole of order d. By the singularity correspondence theorem, $h_{D(0)}(w)$ has a unique singularity at the origin, which is also a pole of order d. Therefore, the same is true for $h_{D(t)}(w)$ for all t. Hence, f_t also has the only singularity, a pole at infinity of order d. This implies that f_t is a polynomial of degree d.

Richardson showed that $f(\zeta)$ is a polynomial of degree d if and only if $M_k(D) = 0$ for $k \geq d$, and $M_{d-1} \neq 0$.

For a domain with uniformization map $f(\zeta) = a_1\zeta + a_2\zeta^2 + \cdots + a_d\zeta^d$, the moments are

$$M_k(D) = \sum_{m_1, \ldots, m_{k+1} \geq 1, \sum m_i \leq d} m_1 a_{m_1} a_{m_2} \cdots a_{m_{k+1}} \bar{a}_{m_1 + \cdots + m_{k+1}}. \tag{3.2}$$

This is a system of equations for finding a_j when the moments M_k are given. It is known that if $df/d\zeta \neq 0$ for $|\zeta| \leq 1$ then this system is nondegenerate [10].

Let us write down the equations for the case when all singularities of the Cauchy transform are simple poles. Then

$$h_D(z) = \sum_{j=1}^{m} \frac{A_j}{z - B_j}. \tag{3.3}$$

Let us look for f in the form

$$f(\zeta) = \sum_{j=1}^{m} \frac{C_j \zeta}{1 - E_j \zeta}. \tag{3.4}$$

This form is chosen to ensure that $f(0) = 0$ (we assume $0 \in D$) and to allow the possibility of a pole at infinity, i.e. $E_j = 0$ for some j. We have

$$\overline{f(1/\bar{\zeta})} = \sum_{j=1}^{m} \frac{\bar{C}_j}{\zeta - \bar{E}_j}.$$

Equations. 1. The poles of $\overline{f(1/\bar{\zeta})}$ coincide with the poles of $h_D(f(\zeta))$:

$$f(E_j) = B_j, \quad j = 1, 2, \ldots, m. \tag{3.5}$$

2. The residues of the functions $\overline{f(1/\bar{\zeta})}$ and $h_D(f(\zeta))$ coincide:

$$A_j = f'(\bar{E}_j)\overline{C}_j, \quad j = 1, 2, \ldots, m. \tag{3.6}$$

The system of equations (3.5), (3.6) contains $2m$ equations with $2m$ unknowns C_j, E_j. However, these equations are not independent. Indeed, if C_j, E_j is a solution then $C_j e^{i\theta}$, $E_j e^{i\theta}$ is also a solution, since these sets of parameters correspond to the same domain.

REMARK. Let f^s be a smooth family of solutions of the above system, and let f^{s_0} map the unit disk conformally onto a domain D. Then so does f^s for any s. Indeed, the domains $D_s = f_s(K)$ have the same Cauchy transform by definition, hence $D_s \equiv D$ by the local uniqueness theorem.

3.7. Examples. 1. (P. Ya. Polubarinova-Kochina, P. P. Kufarev, **[11]**, **[12]**). Let $D(0)$ be the circle of a radius R centered at the point $(a, 0)$, and let the only source be situated at the origin and work at a rate q. Then

$$h_{D(0)}(w) = \frac{R^2}{w-a}, \qquad h_{D(t)}(w) = \frac{qt}{\pi w} + \frac{R^2}{w-a}.$$

The function $h_{D(t)}$ has two simple poles at 0 and a, hence f_t has two simple poles at ∞ and some point α_t, so

$$f_t(\zeta) = \frac{\beta_t \zeta}{1 - \alpha_t \zeta} + \gamma_t \zeta. \tag{3.7}$$

Thus, $\partial D(t)$ is an algebraic curve of degree no larger than four.

Due to the symmetry with respect to the horizontal axis, we can assume that α_t, β_t, γ_t are real numbers, and therefore $\overline{f_t(\bar{z})} = f_t(z)$. Let $\alpha_t > 0$. Then equations (3.5), (3.6) have the form

$$\frac{\beta_t \alpha_t}{1 - \alpha_t^2} + \gamma_t \alpha_t = a,$$

$$\frac{\beta_t^2}{(1 - \alpha_t^2)^2} + \gamma_t \beta_t = R^2,$$

$$(\beta_t + \gamma_t)\gamma_t = \frac{qt}{\pi}.$$

From these equations after some calculations we obtain

$$\gamma_t = \frac{1}{2}\left(\frac{a}{\alpha_t} - \frac{\alpha_t}{a}\left(R^2 - \frac{qt}{\pi}\right)\right),$$

$$\beta_t = \frac{1 - \alpha_t^2}{2}\left(\frac{a}{\alpha_t} + \frac{\alpha_t}{a}\left(R^2 - \frac{qt}{\pi}\right)\right),$$

and α_t can be obtained from the bicubic equation

$$\left(R^2 - \frac{qt}{\pi}\right)^2 \alpha_t^6 - \left(2a^2R^2 + 2a^2\frac{qt}{\pi} + a^4\right)\alpha_t^2 + 2a^4 = 0. \tag{3.8}$$

Considering the case $t = 0$, we find that α_t is the middle root of this equation.

Let $q < 0$. Then at some time the evolving boundary develops a semicubic cusp singularity which is interpreted as a rush of water into the oil-extracting well. It is interesting to calculate how much oil will be extracted by that time. Equating the discriminant of (3.8) to zero, we can find both the time of the rush and the quantity of extracted oil. For instance, if the distance from the well to the center of the oil spot is one half of its radius, then the amount of extracted oil is approximately equal to 10% of the total volume of the spot. In order to secure extraction of one half of all oil before the rush, one needs to drill the well as close to the center of the circle as 0.23 of its radius.

The boundary of the fluid domain is a curve of order four. Its equation is $(x^2+y^2)^2+2\alpha_t a(x^2+y^2)x+K_t(x^2+y^2)+L_t x+M_t=0$, where K_t, L_t, and M_t are some coefficients, which can be calculated by explicit formulas. It is easy to check that this curve can be obtained from an ellipse by an inversion.

2. Suppose that the initial domain $D(0)$ is the image of the unit disk under the mapping

$$f_0(\zeta) = \frac{a}{\pi}\log\frac{1+\alpha_0\zeta}{1-\alpha_0\zeta}, \quad 0<\alpha_0<1, \quad a>0.$$

This domain has an oval-like shape (Figure 6), and it is more prolate the larger the number α_0. When $\alpha_0 = 1$, the function f_0 maps the unit disk onto the strip $-\frac{a}{2} < y < \frac{a}{2}$. Consider the evolution $D(t)$ of this domain in time produced by injecting the fluid from a source of rate q at $z=0$. Let us determine the conformal map f_t of the unit disk onto $D(t)$.

By the correspondence of singularities, $h_{D(0)}(z)$ has two logarithmic branch points. Because the domain is symmetric, these points are real and symmetric with respect to the origin. Let their coordinates be F_0 and $-F_0$, and let $F_0>0$. The function $h_{D(0)}(z)$ has no other singular points, therefore

$$h_{D(0)}(z) = \frac{a}{\pi}\log\frac{z+F_0}{z-F_0},$$

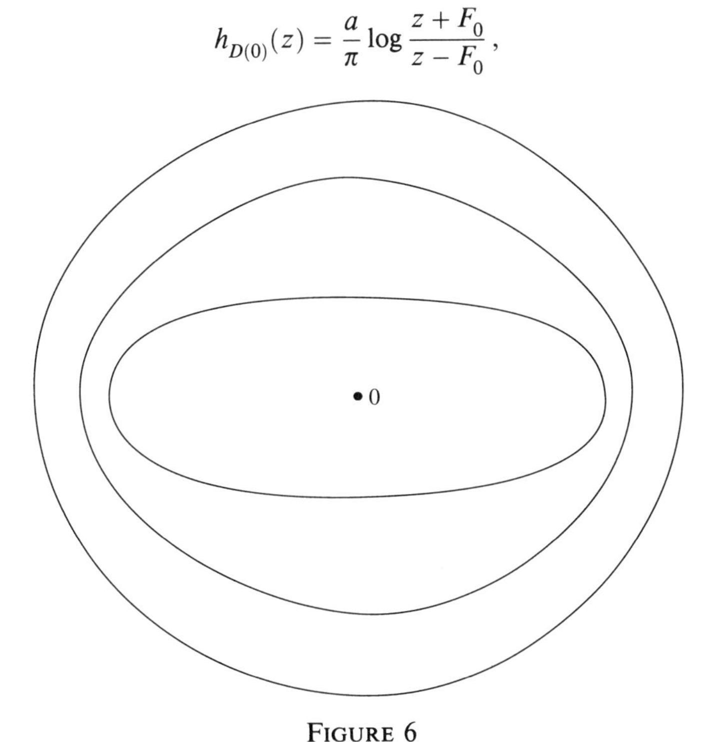

FIGURE 6

where

$$F_0 = f_0(\alpha_0) = \frac{a}{\pi} \log \frac{1+\alpha_0^2}{1-\alpha_0^2}$$

(the logarithms in f_0 and $h_{D(0)}$ must occur with the same coefficient because of the singularity correspondence). Consequently, by Richardson's theorem,

$$h_{D(t)}(z) = \frac{a}{\pi} \log \frac{z+F_0}{z-F_0} + \frac{qt}{\pi z}.$$

This implies that the analytic continuation of $f_t(\zeta)$ outside the unit circle has two logarithmic branch points, say, $1/\alpha_t$ and $-1/\alpha_t$, and a simple pole at infinity ($\alpha_t > 0$, $\alpha_t = \alpha_0$ when $t = 0$). Hence, the uniformization mapping has the form

$$f_t(\zeta) = \frac{a}{\pi} \log \frac{1+\alpha_t \zeta}{1-\alpha_t \zeta} + \gamma_t \zeta.$$

The parameters α_t and γ_t are found from the equations

$$F_0 = f_t(\alpha_t) = \frac{a}{\pi} \log \frac{1+\alpha_t^2}{1-\alpha_t^2} + \gamma_t \alpha_t$$

—correspondence of the logarithmic branch points;

$$\gamma_t = \frac{qt}{\pi f_t'(0)} = \frac{qt}{\pi(\gamma_t + 2a\alpha_t/\pi)}$$

—coincidence of the residues of f_t and $h_{D(t)}$ at the simple poles. From these equations we obtain an equation for determining α_t

$$\frac{a}{\pi} \log \frac{1+\alpha_t^2}{1-\alpha_t^2} + \alpha_t \sqrt{\frac{\alpha_t^2 a^2}{\pi^2} - \frac{qt}{\pi}} - \alpha_t^2 \frac{a}{\pi} = F_0,$$

and γ_t is expressed in terms of α_t as follows:

$$\gamma_t = \frac{-\alpha_t a}{\pi} + \sqrt{\frac{\alpha_t^2 a^2}{\pi^2} - \frac{qt}{\pi}}.$$

Problems. 1. Show that the boundary of an algebraic domain D whose uniformization map is a polynomial is an algebraic curve of degree $2 \deg D$.

2. Are the following domains algebraic: a) $x^4 + y^4 \leq 1$, b) $x^2 + 2y^2 \leq 1$?

3. Consider a domain with uniformization map $a_0\zeta + b_0\zeta^2$ **[11]**. Find the evolution of this domain produced by extraction of oil from the origin at a rate q. Determine the time of cusp development on the boundary and the amount of oil extracted by this time. For what ratio $|a_0/b_0|$ will this amount equal one half of all the oil?

4. Consider the asymptotics of the solution of Example 1 when $R \to \infty$, $a = R - b$. This corresponds to extraction from a half-plane saturated with oil (Figure 7). Show that the quantity of oil extracted before water

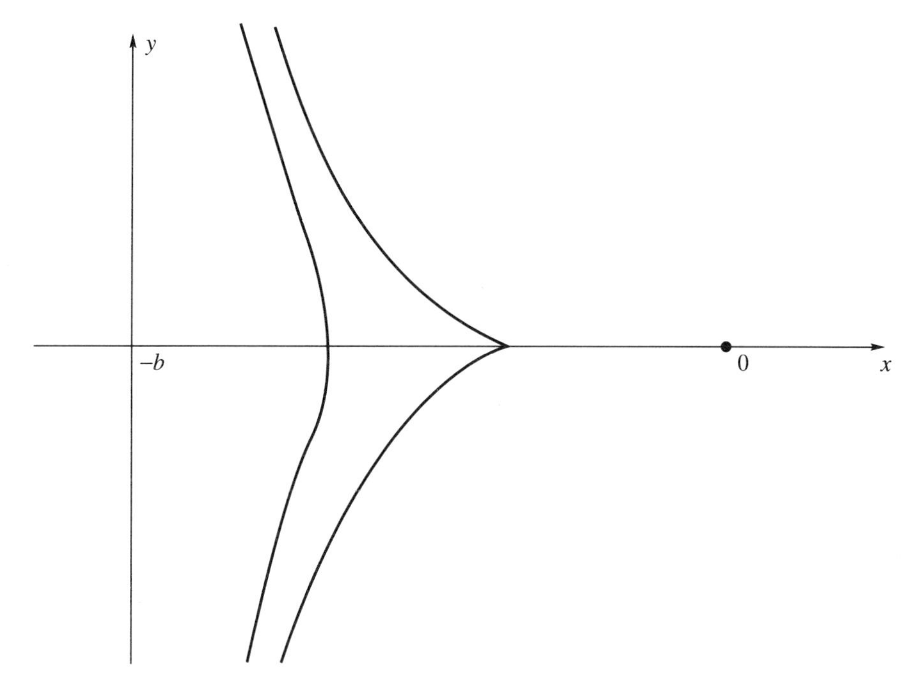

FIGURE 7

rushes into the well (formation of a cusp) equals $\pi b^2/3$, i.e. three times less than in case of extraction from a circular oil spot of radius b (Figure 7). Thus, in this case we have a surprising effect: extension of the initial domain speeds up the water rush.

5. Consider the asymptotics of Example 2 as $\alpha_0 \to 0$. It corresponds to the problem of extraction from a strip (Figure 8). Find the time of cusp development.

6. (P. Ya. Polubarinova-Kochina, P. P. Kufarev). Describe the dynamics

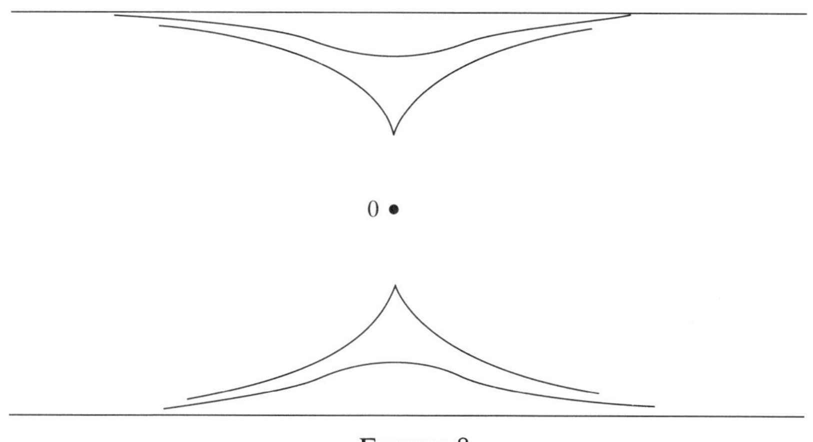

FIGURE 8

of a round oil spot under extraction through N sinks situated at vertices of a regular N-gon and working at the same rate.

7. A round initial domain evolves under injection of fluid from a source of a constant rate that moves uniformly around some circle. Find the shape of the fluid domain at the time the source has made a full round.

4. Contraction of a Gas Bubble

4.1. Formulation of the problem. Consider a bounded simply connected gas domain whose exterior is filled with a fluid (Figure 9). As gas is extracted, the gas-fluid boundary contracts. It is interesting to study the dynamics of this contraction. This was done first in paper **[14]**, and most of the results of this section are contained there.

Let S be the initial area of the gas domain, and let q be the rate of extraction. Then all the gas will be exhausted at the time $t^* = S/q$. In many cases the gas-fluid interface contracts to a point as $t \to t^*$ and is approximately elliptic when t is close to t^*. It is surprising that the position of this point as well as the principal directions and eccentricity of the asymptotic ellipse can be easily found if the initial gas domain is given.

In the first approximation, evolution of the boundary does not depend on the positions of the gas-extracting wells. This is why, from the practical point of view, it makes sense to drill a well at the contraction point: otherwise the

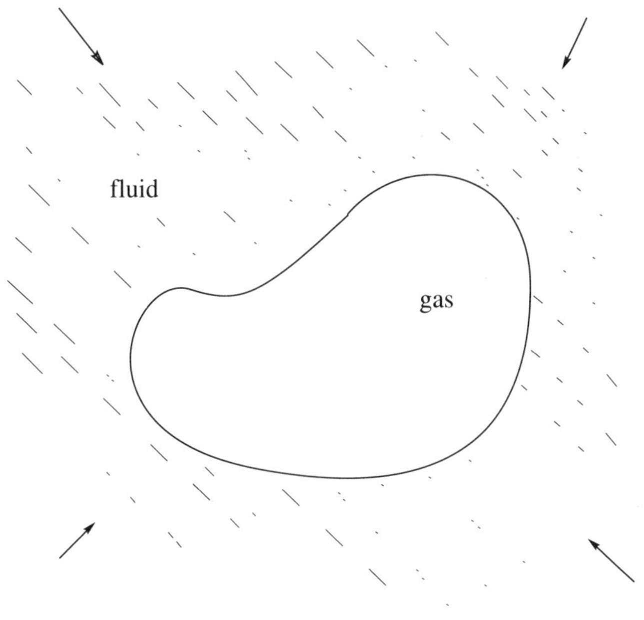

FIGURE 9

water that displaces gas will rush into the well before all the gas has been extracted (see [5]).

Mathematical model. The extracted gas is replaced by water. It is assumed that water is injected from a source situated at infinity. In all other respects, the model is similar to the one presented in Chapter 1. Namely, let $D(t)$ be the gas domain at a time t, and let $\overline{D(t)}$ be its complement. Then at any time t one defines a potential fluid velocity field $v_t = \operatorname{grad}\Phi_t$, the velocity potential Φ_t being the solution of Laplace's equation

$$\Delta\Phi_t = 0 \quad \text{in } \overline{D(t)}$$

that vanishes on the boundary:

$$\Phi_t \,|_{\partial D(t)} = 0$$

and has the logarithmic asymptotics at infinity:

$$\Phi_t(z) \sim -\frac{q}{2\pi}\log|z|, \qquad z \to \infty.$$

For the sake of convenience let us assume that $\Phi_t = 0$ inside $D(t)$ as well as on the boundary.

The law of motion is: boundary points, like all other points of $\overline{D(t)}$, move with a velocity of $\operatorname{grad}\Phi_t$.

REMARK. One can assume that the domain $D(t)$ consists of several simply connected components. This describes simultaneous contraction of a number of gas bubbles.

For any initial domain $D(0)$ with a smooth boundary, the evolution of the boundary is well defined on some time interval $[0, \tau]$, and the boundaries $\partial D(t)$ for $0 < t \le \tau$ are regular analytic curves (see [13]).

4.2. The inclusion property.

LEMMA. *When contracting, the domain monotonically decreases*: for $t_1 < t_2$ one has $D(t_1) \supset D(t_2)$.

Indeed, by the maximum principle, $\Phi_t \le 0$ in $\overline{D(t)}$, so the gradient of Φ_t is directed inwards throughout the boundary. This implies that the boundary moves inwards everywhere.

COROLLARY. *The intersection* $\bigcap_{t<t_0} D(t)$ *coincides with the closure of* $D(t_0)$.

Now suppose that a solution of problem 4.1 is defined on the open time interval $(0, \tau)$, and the intersection $\bigcap_{t<\tau} D(t)$ is the closure of a simply connected domain. Then this solution can be extended to a larger time interval by considering further contraction of the boundary of this domain.

Thus, one can distinguish two kinds of initial domains:

1. Domains that remain connected in the course of contraction and vanish as soon as all the gas has been extracted.

2. Domains that cease to be connected in the process of contraction.

In this chapter we will examine domains of the first kind. Domains of the second kind are considered in Chapter 6.

4.3. Contraction of a convex domain.

THEOREM. *A convex domain remains convex when contracting.*

COROLLARY. *A convex domain remains connected and contracts completely when all the gas has been extracted.*

LEMMA. *Let D be a convex domain, Φ be the velocity potential outside D defined in* § 4.1, *and let $\frac{\partial^2\Phi}{\partial s^2}(P)$ be the second derivative along the tangent line to the level curve of Φ at $P \in \overline{D}$. Then $\frac{\partial^2\Phi}{\partial s^2} < 0$ in $\overline{D}$, and it can vanish only on the boundary. At points of the boundary where it vanishes, the normal derivative $\frac{\partial}{\partial n}\left(\frac{\partial^2\Phi}{\partial s^2}\right)$ is positive.*

PROOF. Introduce an auxiliary function $\Omega(P) = \frac{1}{|\operatorname{grad}\Phi(P)|^2}\frac{\partial^2\Phi}{\partial s^2}(P)$ that is defined in $\overline{D}$. It is harmonic in $\overline{D}$ and regular at infinity. The simplest way to show it, is to establish that $\Omega = 2\operatorname{Re}\left(\frac{d}{dz}\left(\frac{1}{\omega(z)}\right)\right)$, where $\omega(z)$ is the complex velocity of the flow $\frac{\partial\Phi}{\partial x} - i\frac{\partial\Phi}{\partial y}$. The sign of the function $\Omega(P)$ coincides with that of $\frac{\partial^2\Phi}{\partial s^2}(P)$ and is opposite to that of the curvature of the level curve of the potential at P. Since ∂D is a convex level curve of the potential, $\frac{\partial^2\Phi}{\partial s^2}$ and Ω are nonpositive along ∂D. By the maximum principle, from this it follows that Ω (and hence, $\frac{\partial^2\Phi}{\partial s^2}$) is negative inside $\overline{D}$. Moreover, the strong maximum principle of Hopf **[15]** claims that the maximal value of a non-constant harmonic function cannot be attained at a point of the boundary where its normal derivative vanishes. This implies $\frac{\partial\Omega(P)}{\partial n} > 0$ for $P \in \partial D$ such that $\Omega(P) = 0$. Consequently, $\frac{\partial}{\partial n}\left(\frac{\partial^2\Phi}{\partial s^2}(P)\right) = |\operatorname{grad}\Phi(P)|^2\frac{\partial\Omega(P)}{\partial n} > 0$ for $P \in \partial D$ such that $\frac{\partial^2\Phi}{\partial s^2}(P) = 0$.

COROLLARY. *For a convex domain D, all level curves of the potential Φ outside D are strictly convex.*

PROOF OF THE THEOREM. Let τ be the time when the gas domain ceases to be convex. Suppose that $\tau \neq t^*$, and the violation of convexity at $t > \tau$ takes place in a neighborhood of a point $Q \in \partial D(\tau)$. It is clear that the curvature of the boundary at this point equals zero (Figure 10, p. 24). It is seen from this picture that the velocity v of the boundary in a neighborhood of Q is a concave function of the natural parameter s on $\partial D(\tau)$:

$$\frac{\partial^2 v}{\partial s^2}(Q) = \frac{\partial^2}{\partial s^2}\left(\frac{\partial\Phi}{\partial n}\right) \leq 0, \quad \text{i.e.} \quad \frac{\partial}{\partial n}\left(\frac{\partial^2\Phi}{\partial s^2}\right)(Q) \leq 0,$$

which contradicts the above lemma.

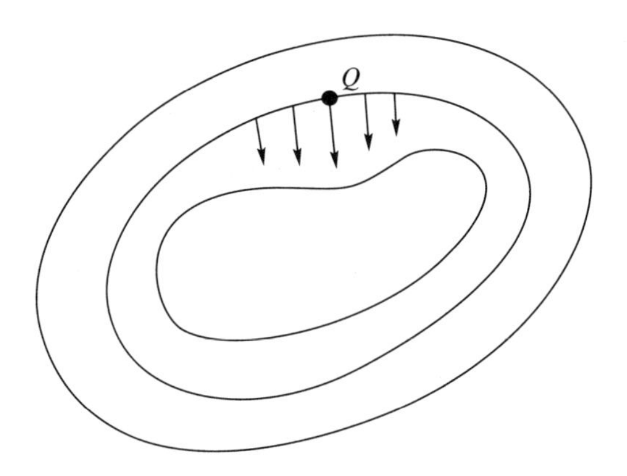

FIGURE 10

REMARK. We have also proved that the domains $D(t)$ for $t > 0$ are strictly convex.

4.4. Contraction points.

DEFINITION. A point is called a *contraction point* of the domain $D(0)$ (or contour $\partial D(0)$) if it belongs to the gas domain $D(t)$ at all times $t \in [0, t^*)$, t^* being the time when all the gas is extracted.

THE MAIN THEOREM. *Let the domain $D(t)$ remain connected in the course of contraction, and contract completely when all the gas has been extracted. Consider its gravity potential,*

$$\Pi_{D(t)}(w) = \frac{1}{2\pi} \int_{D(t)} \log|z - w|\, dx\, dy.$$

1. *The gravity potential $\Pi_{D(t)}$ changes by a constant in the course of contraction*: $\Pi_{D(0)} - \Pi_{D(t)} = \text{const}(t)$ in $D(t)$.

2. Any smooth family of domains whose area depends linearly on time and whose potential changes by a constant in time, is a solution of the contraction problem.

3. A contraction point is unique.

4. The minimal value of the gravity potential of a domain is attained at its contraction point, and only there.

The proof of this theorem is contained in §§4.5–4.8.

It is convenient to look for the contraction point of a domain among its "geometric centers".

DEFINITION. Let us say that a point P is a *geometric center* of a domain D if this point satisfies the property

$$\int_0^{2\pi} \mathbf{r}_P(\theta) d\theta = 0, \tag{4.1}$$

where $\mathbf{r}_P(\theta)$ is the radius-vector of a boundary point of D whose slope angle is θ (Figure 11).

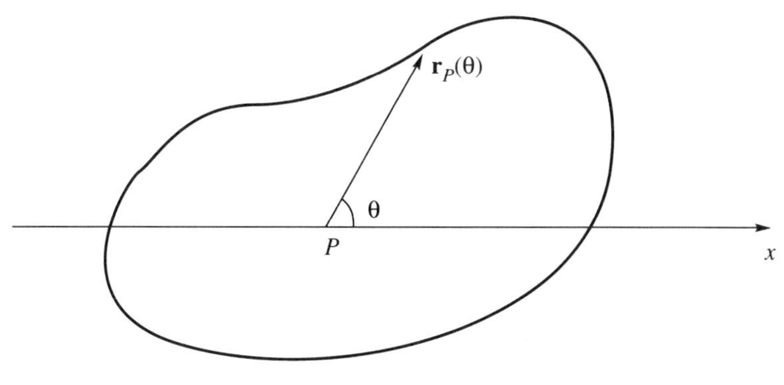

FIGURE 11

One can easily show the following.

LEMMA. *The vector-valued function* $a(P) = \int_0^{2\pi} \mathbf{r}_P(\theta)d\theta$ *is proportional to the gradient of the gravity potential* $\Pi_{D(0)}$.

COROLLARIES. 1. *Geometric centers are critical points of the gravity potential.*

2. *The contraction point of a domain is a geometric center of this domain.*

As time approaches the moment of full extraction, the contracting domain acquires an approximately elliptic shape. The principal axes of the limiting ellipse are directed along the eigenvectors of the matrix of second derivatives of the gravity potential at the contraction point. The lengths of the principal axes are proportional to the eigenvalues of this matrix.

The technical tool that enables us to obtain these results is a suitable generalization of Richardson's theorem.

4.5. An analogue of Richardson's theorem.

DEFINITION. Let us say that a function $u(z)$ defined on the complex plane is *regular at infinity* if the function $u(1/z)$ is defined and continuously differentiable in a neighborhood of $z = 0$.

Consider the contracting domain $D(t)$ and an arbitrary domain G that contains $D(t)$, for instance, a sufficiently large disk.

THEOREM. *For any function* $u(z)$ *that is regular at infinity,*

$$\frac{d}{dt}\int_{G\setminus D(t)} u\,dx\,dy = qu(\infty) - \int_{\overline{D(t)}} \Phi_t\,\Delta u\,dx\,dy. \tag{4.2}$$

REMARK. The integral in the right-hand side of (4.2) converges because $\Delta u(z) = O(1/|z|^4)$ and $\Phi_t(z) \sim -\frac{q}{2\pi}\log|z|$ as $z \to \infty$.

COROLLARY. *For a harmonic, regular at infinity function* $u(z)$,

$$\frac{d}{dt}\int_{G\setminus D(t)} u\,dx\,dy = qu(\infty). \tag{4.3}$$

EXAMPLE.

$$\frac{d}{dt}\int_{G\setminus D(t)} z^{-k}\,dx\,dy = \begin{cases} q, & k=0, \\ 0, & k>0. \end{cases} \tag{4.4}$$

For $k = 0$, this means that the rate of decrease of the area of the gas domain equals the rate of injection of water. For $k \geq 3$, one can take for G the entire plane, and integrate over the complement of $D(t)$, i.e. over the fluid domain. In this case, the analogy with Richardson's theorem is most apparent.

Properties of integrals (4.4) are similar to those of Richardson's integrals.

THEOREM (on local uniqueness). *For a nontrivial continuous family of domains $D(s)$ at least one of the integrals $\int_{G\setminus D(t)} z^{-k}\,dx\,dy$ is a nonconstant function of s.*

THEOREM (on algebraic solutions).

1. *If the complement of the initial domain $D(0)$ is the image of the unit disk under a rational conformal mapping, then so is the complement of $D(t)$ for all $t \in [0, t^*)$.*

2. *If the derivative of a conformal mapping of the unit disk into the complement of $D(0)$ is a rational function, then so is the corresponding derivative for $D(t)$ when $t > 0$.*

Thus, the property of the complement of $D(t)$ being algebraic or abelian is preserved in the course of contraction. The coefficients of the uniformization mapping can be found from a system of nonlinear equations similar to (3.5), (3.6).

PROOF OF FORMULA (4.2).

$$\begin{aligned} \frac{d}{dt}\int_{G\setminus D(t)} u\,dx\,dy &= \int_{\partial D(t)} u\frac{\partial \Phi_t}{\partial n}\,dl = \int_{\partial D(t)}\left(u\frac{\partial \Phi_t}{\partial n} - \Phi_t\frac{\partial u}{\partial n}\right)dl \\ &= \int_{\partial K_R}\left(u\frac{\partial \Phi_t}{\partial n} - \Phi_t\frac{\partial u}{\partial n}\right)dl \\ &\quad + \int_{K_R\setminus D(t)} (u\Delta\Phi - \Phi\Delta u)\,dx\,dy, \end{aligned}$$

where K_R is a circle of radius R, centered at the origin, that contains the domain $D(t)$. The value of this expression is independent of R. Taking into account the asymptotics

$$u\,|_{\partial K_R} = u(\infty) + O(1/R), \qquad \left.\frac{\partial u}{\partial n}\right|_{\partial K_R} = O(1/R^2),$$

$$\left.\Phi_t\right|_{\partial K_R} = \frac{q}{2\pi}\log R + O(1), \qquad \left.\frac{\partial \Phi_t}{\partial n}\right|_{\partial K_R} = \frac{q}{2\pi R} + O(1/R^2), \quad R \to \infty,$$

and also the harmonicity of Φ in $\overline{D(t)}$, we obtain (4.2) in the limit as $R \to \infty$.

4.6. Dynamics of the gravity potential. Recall that the gravity potential of a domain D is defined by formula (2.3):

$$\Pi_D(\xi, \eta) = \frac{1}{4\pi}\int_D \log((x-\xi)^2 + (y-\eta)^2)\, dx\, dy.$$

THEOREM. *The gravity potential of a contracting domain changes in time as follows:*

$$\frac{d}{dt}\Pi_{D(t)}(\xi, \eta) = C(t) + \Phi_t(\xi, \eta), \tag{4.5}$$

where Φ_t is the velocity potential of the fluid flow.

PROOF. In order to apply Theorem 4.5, let us rewrite the left-hand side of (4.5) so that the integration be over the domain $G\backslash D(t)$, and the integrand be regular at infinity:

$$\begin{aligned}\frac{d}{dt}\Pi_{D(t)}(\xi, \eta) &= -\frac{d}{dt}\frac{1}{4\pi}\int_{G\backslash D(t)} \log((x-\xi)^2 + (y-\eta)^2)\, dx\, dy \\ &= -\frac{d}{dt}\frac{1}{4\pi}\int_{G\backslash D(t)} \log\frac{((x-\xi)^2 + (y-\eta)^2)}{((x-\xi_0)^2 + (y-\eta_0)^2)}\, dx\, dy \\ &\quad - \frac{d}{dt}\frac{1}{4\pi}\int_{G\backslash D(t)} \log((x-\xi_0)^2 + (y-\eta_0)^2)\, dx\, dy,\end{aligned} \tag{4.6}$$

(ξ_0, η_0) being an arbitrary point in $D(t)$. The function

$$u(x, y) = \frac{1}{4\pi}\int_{G\backslash D(t)} \log\frac{((x-\xi)^2 + (y-\eta)^2)}{((x-\xi_0)^2 + (y-\eta_0)^2)}\, dx\, dy$$

satisfies Poisson's equation $\Delta u = \delta(x-\xi, y-\eta)$ in $\overline{D(t)}$, $\delta(x, y)$ being Dirac's function. Besides, this function is regular and vanishes at infinity. Therefore, by Theorem 4.5, the first term in (4.6) equals $\Phi_t(\xi, \eta)$. The second term does not depend on ξ, η. Denoting it by $C(t)$, we come to (4.5).

4.7. The gravity potential as the solution of a boundary value problem. It follows from the definition of the gravity potential that it is continuously differentiable on the whole plane, and satisfies the Poisson's equation

$$\Delta\Pi_D(x, y) = \chi_D(x, y) = \begin{cases} 1, & (x, y) \in D, \\ 0, & (x, y) \neq D, \end{cases} \tag{4.7}$$

with the boundary condition at infinity

$$\Pi_D(x, y) \sim \frac{S}{4\pi}\log(x^2 + y^2), \quad (x, y) \to \infty, \tag{4.8}$$

S being the area of D. It determines the gravity potential uniquely, up to an additive constant.

4.8. Proof of the main theorem. Let us integrate both sides of identity (4.5) from 0 to t. We get

$$\Pi_{D(0)}(\xi, \eta) - \Pi_{D(t)}(\xi, \eta) = -\int_0^t C(\tau)d\tau - \int_0^t \Phi_\tau(\xi, \eta)d\tau. \tag{4.9}$$

Set $K(t) = -\int_0^t C(t)dt$. Since $\Phi_t(\xi, \eta) \leq 0$, and it equals zero if and only if (ξ, η) belongs to the closure of the gas domain, $[D(t)]$, identity (4.9) yields

1) (statement 1 of the main theorem):

$$\Pi_{D(0)}(\xi, \eta) - \Pi_{D(t)}(\xi, \eta) = K(t), \quad (\xi, \eta) \in [D(t)]; \tag{4.10}$$

2)

$$\Pi_{D(0)}(\xi, \eta) - \Pi_{D(t)}(\xi, \eta) > K(t), \quad (\xi, \eta) \notin [D(t)]; \tag{4.11}$$

3) The difference $\Pi_{D(0)}(\xi, \eta) - \Pi_{D(t)}(\xi, \eta) - K(t)$ monotonically increases as a function of t for fixed ξ and η.

Set now $t^* = S/q$. As $t \to t^*$, the area of the contracting domain tends to zero, so $\Pi_{D(t)}(\xi, \eta)$ goes to zero as well.

If (ξ, η) is a contraction point then it must belong to $[D(t)]$ for all t; therefore, as $t \to t^*$, (4.10) turns into $\Pi_{D(0)}(\xi, \eta) = K(t^*)$. If (ξ, η) is not a contraction point then there is a τ such that $(\xi, \eta) \notin [D(t)]$ for $t > \tau$, and taking the limit $t \to t^*$ in (4.11), one gets $\Pi_{D(0)}(\xi, \eta) > K(t^*)$. This proves statement 4 of the theorem.

Let us give a proof of statement 2 of the main theorem. Let $D(t)$ be the solution of the contraction problem, and let $D'(t)$ be another family of domains of area $S - qt$ whose potential is independent of time up to an additive constant, and such that $D(0) = D'(0) = D$. Consider, for some time τ, the family of domains $D(\tau, s)$ that are obtained from $D'(\tau - s)$ in the course of contraction during time s, $s \in [0, \tau]$. The domains $D(\tau, s)$ have area $S - q\tau$, and their potentials may differ only by an additive constant inside their intersection. This implies that moments (4.4) of $D(\tau, s)$ do not depend on s. By the local uniqueness theorem, $D(\tau, s)$ does not depend on s either. Hence $D(\tau, \tau) = D(\tau, 0)$. But $D(\tau, \tau) = D(\tau)$, $D(\tau, 0) = D'(\tau)$. Thus, $D(t) = D'(t)$ for all t.

Let us now prove statement 3, the uniqueness of the contraction point. Consider the set of all contraction points. This set is

1) simply connected, since it is the intersection of the simply connected gas domains;

2) bounded;

3) analytic, since it is the set of solutions of the equation $\Pi_{D(0)} = K(t^*)$, and Π_D is a real analytic function in D (because it satisfies Poisson's equation $\Delta\Pi_D = 1$).

Therefore, the set of contraction points must either consist of a single point or be an analytic curve. A bounded analytic curve has the topology of a finite graph. This graph has to be a tree because it is automatically simply connected. A tree always has a vertex that has a unique outgoing edge. But there are no analytic curves with such local topology [16]. Thus, the contraction set is a point.

REMARK. Taking the limit $t \to t^*$ in (4.9), we obtain the following expression of the gravity potential of a domain in terms of the function Φ_t:

$$\Pi_{D(0)}(\xi, \eta) = K - \int_0^{t^*} \Phi_t(\xi, \eta) dt, \tag{4.12}$$

where K is some constant.

This formula implies statement 4 of the main theorem.

4.9. Self-similar solutions.

THEOREM.

1. *A decreasing family of similar ellipses with a common center is a solution of the contraction problem.*
2. *Any decreasing family of similar domains that is a solution of the contraction problem is a family of ellipses with a common center.*

PROOF. 1. Newton discovered that the gravity potential inside an ellipse changes by a constant as the ellipse expands so that its shape and center stay unchanged. Therefore, by statement 2 of the main theorem, a family of similar concentric ellipses is a solution of the contraction problem.

2. The proof is based on an application of a theorem due to M. Sakai, [17]: if D is a simply connected domain and $\overline{D}$ is its complement, and if $\int_{\overline{D}} z^{-k} dxdy = 0$ for $k \geq 3$ then D is an ellipse.

Let $D(t)$ be a family of domains homothetic with respect to the origin that is a solution of the contraction problem. The domain $D(t)$ is obtained from $D(0)$ by a homothetic contraction. The coefficient of contraction has to be $\lambda(t) = \sqrt{S/(S - qt)}$, therefore

$$\int_{\overline{D(t)}} z^{-k} dxdy = \lambda(t)^{2-k} \int_{\overline{D(0)}} z^{-k} dxdy, \quad k \geq 3,$$

where $\overline{D(t)}$ is the complement of $D(t)$. On the other hand, $\int_{\overline{D(t)}} z^{-k} dxdy =$ const because of (4.4). Hence,

$$\int_{\overline{D(0)}} z^{-k} dxdy = 0, \quad k \geq 3. \tag{4.13}$$

Therefore, $D(0)$ is an ellipse. But we know that in the course of evolution, an ellipse contracts homothetically with respect to its center. This completes the proof.

REMARK. The gravity potential of the ellipse $\frac{x^2}{a^2} + \frac{y^2}{b^2} = 1$ inside itself can be easily calculated:

$$\Pi = \frac{1}{2}\left(\frac{a}{a+b}x^2 + \frac{b}{a+b}y^2\right) + C(a, b). \tag{4.14}$$

In particular, the inner gravity potential of the disk of a radius R centered at the origin equals

$$\Pi = \frac{1}{4}(x^2 + y^2) + C(R). \tag{4.15}$$

4.10. Asymptotics of contraction. Let $D(0)$ be a convex domain that contracts to the origin. All intermediate domains $D(t)$ are also convex. Consider the domain $E(t)$ that is obtained from $D(t)$ by expanding $\lambda(t)$ times, where λ is chosen so that the length of the boundary of $E(t)$ should equal a fixed number l.

THEOREM. *The boundaries of $\{E(t)\}$ uniformly converge to an ellipse. The principal axes of this ellipse are directed along the eigenvectors of the matrix of second derivatives of the gravity potential $\Pi_{D(0)}$ at the origin. The lengths of the principal axes are proportional to the eigenvalues of this matrix.*

SKETCH OF A PROOF. It suffices to show that in any sequence of times $\{t_n\}$ that tends to t^* as $n \to \infty$ there exists a subsequence $\{t_{n_k}\}$ such that the curves $\partial E(t_{n_k})$ uniformly converge to the ellipse defined in the statement of the theorem.

Any sequence of vector-valued functions on an interval, uniformly bounded together with the first derivative, contains a uniformly convergent subsequence. Let $z = z_t(s)$, $s \in [0, l]$, be a parametric equation of the curve $\partial E(t)$, s being the natural parameter on this curve. Since $|z_t(s)| \leq \frac{l}{2}$, $|\frac{dz_t(s)}{ds}| = 1$, and for all t there exists s such that $|z_t(s)| \geq \frac{l}{2\pi}$, we conclude that any sequence of times t_n contains a subsequence $\{t_{n_k}\}$ such that $z_{t_{n_k}}(s)$ converges uniformly on $[0, l]$ to some nonzero function $Z(s)$ as $k \to$ infinity. The equation $z = Z(s)$ defines a closed curve. One can check that this curve bounds a convex domain. Let us denote this domain by E.

Since for $k \geq 3$

$$\int_{\overline{E(t)}} z^{-k} dxdy = \lambda(t)^{2-k} \int_{\overline{D(t)}} z^{-k} dxdy = \lambda(t)^{2-k} \int_{\overline{D(0)}} z^{-k} dxdy,$$

we find that $\int_{\overline{E(t)}} z^{-k} dxdy$ tends to zero as $t \to t^*$, so

$$\int_{\overline{E}} z^{-k} dxdy = 0, \quad k \geq 3.$$

A domain that has this property must be an ellipse (see §4.9). The inner gravity potential of the ellipse E equals

$$\Pi_E(\xi, \eta) = \lim_{t \to t^*} \Pi_{E(t)}(\xi, \eta) = \Pi^{(2)}_{D(0)}(\xi, \eta) + \text{const},$$

where $\Pi^{(2)}_{D(0)}(\xi, \eta)$ are the second order terms in Taylor's expansion of the potential of the initial domain at the origin. Formula (4.14) implies that in an ellipse centered at the origin, the principal axes are directed along the eigenvectors, and their lengths are proportional to eigenvalues of the matrix of coefficients of its gravity potential (which is a sum of a quadratic function and a constant). For the ellipse E, this matrix coincides with the matrix of second derivatives of the gravity potential of $D(0)$.

4.11. Several sources. Consider the problem of contraction of the boundary between fluid and gas in case when gas is injected through several sources. Let q be the rate of the source that is situated at infinity, and let $q_1, \dots, q_n$ be the rates of the sources situated at points $P_1, \dots, P_n$ with coordinates $(x_1, y_1), \dots, (x_n, y_n)$. The velocity potential in this case is defined as the solution of the following boundary value problem:

$$\Delta\Phi_t = 0 \quad \text{in } \overline{D(t)} \backslash \{P_1, \dots, P_n\}, \qquad \Phi_t = 0 \quad \text{on } \partial D(t),$$

$$\Phi_t \sim \frac{q_j}{4\pi} \log((x - x_j)^2 + (y - y_j)^2) \quad \text{as } (x, y) \to (x_j, y_j),$$

$$\Phi_t \sim \frac{q}{4\pi} \log(x^2 + y^2) \quad \text{as } (x, y) \to \infty.$$

In this contraction problem, like in the one considered above, a domain that remains connected in the course of contraction, has a unique contraction point. This point is the minimal value point of a certain function which we choose to call the effective potential. This function changes by an additive constant in the course of evolution. It is formally defined by (4.12), and can be obtained from the gravity potential as follows:

(4.16)

$$\widehat{\Pi}_D(x, y) = \Pi_D(x, y) - \sum_{j=1}^{n} \frac{q_j}{q + q_1 + \cdots + q_n} \frac{S}{4\pi} \log((x - x_j)^2 + (y - y_j)^2).$$

Thus, the effective potential is nothing else but the electric potential of the uniformly charged domain D in the presence of charges of the opposite sign proportional to q_j, situated at points P_j. The proofs of these statements are not principally different from the ones given above.

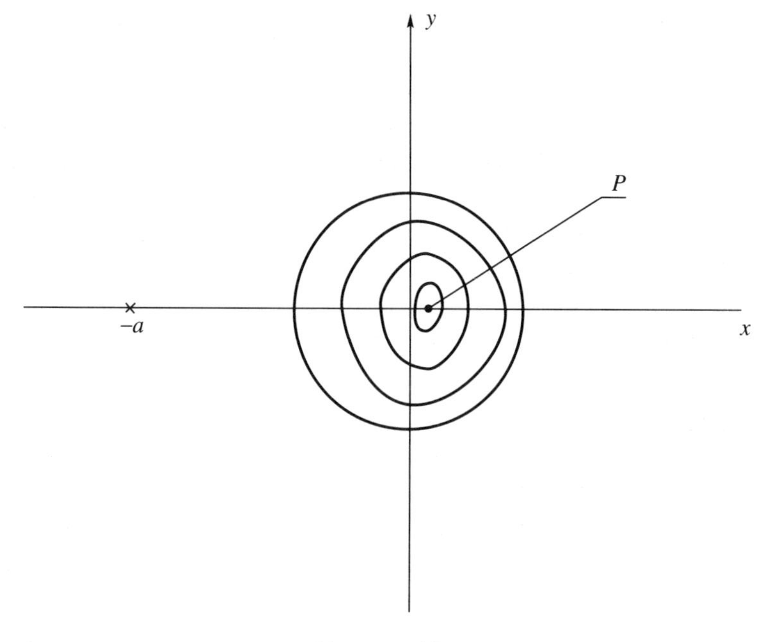

FIGURE 12

EXAMPLE. Suppose that the initial gas domain is the disk of a radius R centered at the origin. Let the only source be situated at a point $(-a, 0)$ (Figure 12). Then the effective potential equals

$$\frac{x^2+y^2}{4} - \frac{R^2}{4}\log((x+a)^2+y^2) + C,$$

and its minimal value is attained at the point $(\frac{-a+\sqrt{a^2+4R^2}}{2}, 0)$. This is exactly the contraction point.

Problems. 1. Consider the process of expansion of a gas domain as a result of extraction of fluid from infinity (Problem 4.1 with $q < 0$). Let $D(t)$, $t > 0$, be the expanding gas domain, and let $\bigcup_{t>0} D(t) = \mathbf{R}^2$. Show that then $D(0)$ has to be an ellipse. Thus, solutions with other initial domains either cease to exist at some finite time t or exist for all t but gas does not eventually fill the whole plane (Howison's theorem, **[18]**).

2. Let the contracting domain D be symmetric with respect to the horizontal axis and remain connected in the course of contraction. Then the function $\Pi_D(x, 0)$, as a function of x, has a unique local minimum that corresponds to the contraction point.

3. Find the contraction point of a round domain in a) a half-plane (Figure 13), b) a strip (Figure 14). In both cases the fluid is injected from infinity. The boundaries are impervious, i.e. $\frac{\partial\Phi}{\partial n} = 0$ along them.

4. Find the ratio of the lengths of the principal axes of the asymptotic ellipse for the contracting domain in the example of §4.11.

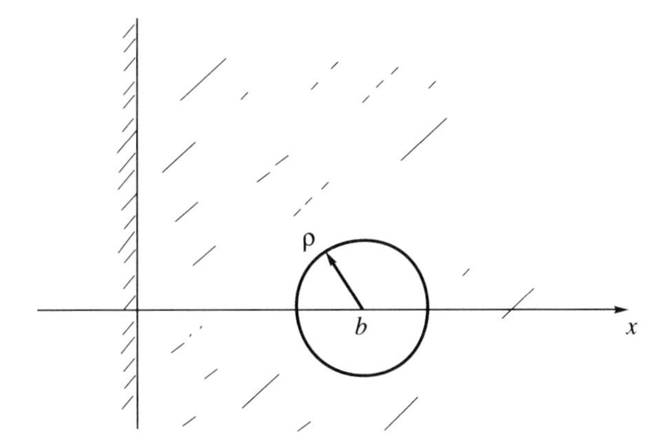

FIGURE 13

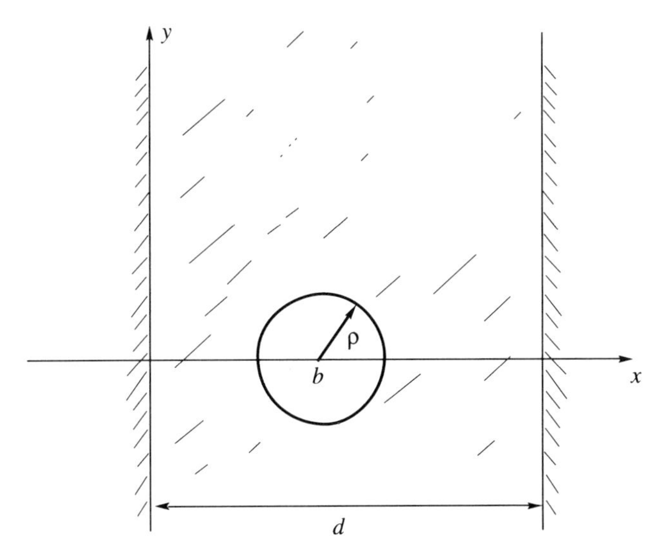

FIGURE 14

5. Evolution of a Multiply Connected Domain

5.1. Statement of the problem. First Let the fluid domain be a bounded domain with g holes (Figure 15). We assume that its complement is filled with gas. Consider the evolution of the domain produced by injection from point sources inside it. Let $D(t)$ be the fluid domain at a time t. Let $z_1, \dots, z_n$ be the complex coordinates of the sources, and denote their rates by $q_1, \dots, q_n$. Let $\Gamma_1, \dots, \Gamma_{g+1}$ be the connected components of the boundary $\partial D(t)$. The pressure is constant along each connected component of the boundary. However, values of the pressure on distinct components may be different. The potential of the velocity field of fluid particles is defined by

$$\Delta\Phi = 0 \quad \text{in } D(t)\backslash\{z_1, \dots, z_n\}, \qquad \Phi \sim \frac{q_j}{2\pi}\log|z - z_j|, \quad z \to z_j,$$

$$\Phi\Big|_{\Gamma_j} = \Phi_j(t), \quad j = 1, \dots, g+1.$$

The boundary velocity equals $v = \frac{\partial\Phi}{\partial n}\big|_{\partial D(t)}$, i.e. the absolute value of the gradient of the velocity potential.

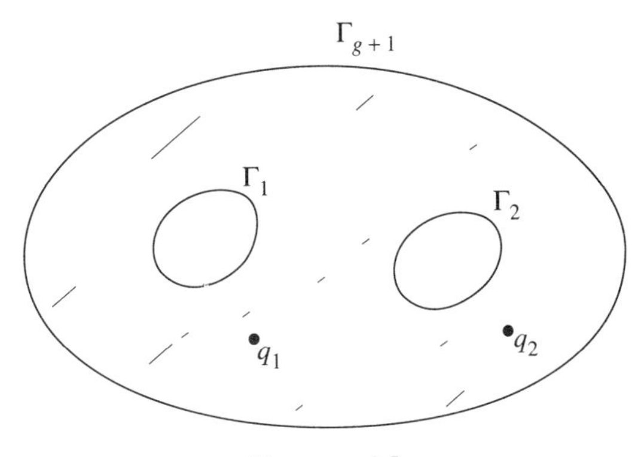

FIGURE 15

5.2. Integrals of motion.

THEOREM (Richardson-Gustafsson). *For every harmonic function u in a neighborhood of $D(t)$*

$$\frac{d}{dt}\int_{D(t)} u\,dx\,dy = \sum_{j=1}^{n} q_j\, u(z_j) + \sum_{j=1}^{g+1} \Phi_j R_j, \tag{5.1}$$

where $R_j = \int_{\Gamma_j}\left(\frac{\partial u}{\partial x}dy - \frac{\partial u}{\partial y}dx\right)$ *is the flux of the gradient of the function* u *through the contour* Γ_j.

REMARK. If the function u is holomorphic in a neighborhood of Γ_j then $R_j = 0$.

Analogously to the simply connected case, let us call the integrals of harmonic functions over a multiply connected domain the moments of this domain.

THEOREM (on the local uniqueness of a domain with given moments). *Let* $D(s)$, $s \in (s_1, s_2)$, *be a smooth family of multiply connected domains, and assume that for every* $s_0 \in (s_1, s_2)$ *and every function* $u(z)$ *harmonic in a neighborhood of* $D(s_0)$, $\int_{D(s)} u\,dx\,dy = \int_{D(s_0)} u\,dx\,dy$ *for* s *close enough to* s_0. *Then* $D(s) = D(t)$ *for any* s, t.

COROLLARY. *The result of injection depends only on the initial domain and the total amounts of fluid injected by the sources. It is independent of the order of their work.*

These statements are proved similarly to their analogues in the simply connected case (see Chapter 2).

5.3. Algebraic solutions. An algebraic domain in the multiply connected case is defined with the help of a Riemann surface with an antiholomorphic involution and a meromorphic differential.

A Riemann surface (a complex curve) is a smooth surface on which, in a neighborhood of every point, a local complex coordinate is defined so that the transition function from one local coordinate to another is a holomorphic function wherever two neighborhoods overlap (see **[19]**).

An antiholomorphic involution of a Riemann surface is a diffeomorphism of the surface onto itself whose square is the identity, and which can be written in local complex coordinates in the form $w = \overline{\phi(z)}$, ϕ being a holomorphic function.

It is said that a Riemann surface Σ of genus g is equipped with a real M-structure if an antiholomorphic involution is so defined on this surface that the set of its fixed points consists of $g+1$ closed contours (ovals). In this case, this set divides the surface into two halves: Σ^+ and Σ^-, which are permuted by the involution.

EXAMPLE. The surface is the Riemann sphere, the involution is a reflection with respect to a circle (inversion).

A meromorphic differential on a surface is a differential 1-form ω that has the form $\phi(z)\,dz$ in local coordinates, ϕ being a meromorphic function.

Let Σ be a Riemann surface with a real M-structure, and let ω be a meromorphic differential on this surface. Denote by $\gamma_1, \dots, \gamma_{g+1}$ the connected components of the set of fixed points of the antiholomorphic involution (the ovals). Assume that $\int_{\gamma_j} \omega = 0$ and that ω has neither zeros nor poles in the hemishell Σ^+. Consider the function $f(z) = \int_{z_0}^{z} \omega$, where z_0 is a point in Σ^+. This function is holomorphic and single-valued on Σ^+. Assume that this function takes different values at different points. Then it defines a conformal mapping of the hemishell Σ^+ onto a plane domain with g holes. Let us call such a domain an *abelian* domain of genus g. Let us define the multiplicity of an abelian domain as the sum of orders of all poles of ω.

If ω is an exact form then the function f extends to a meromorphic function on Σ. In this case let us say that the domain is algebraic of genus g. Define the degree of an algebraic domain as the sum of orders of all poles of the function f. The difference between the multiplicity and the degree of an algebraic domain equals the number of poles of the function f. The boundary of an algebraic domain is always a real algebraic curve that consists of $g+1$ components. Every domain with a finite number of holes can be approximated by an algebraic one as precisely as desired.

THEOREM. *Let $D(t)$ be the evolving fluid domain. Then*:

1. If $D(0)$ is an algebraic domain of genus g and degree d then $D(t)$ is an algebraic domain of genus g and degree no higher than $d+n$.

2. If $D(0)$ is an abelian domain of genus g and multiplicity k then $D(t)$ is an abelian domain of genus g and multiplicity no higher than $k+2n$.

5.4. Riemann's theorem.

THEOREM. *Let D be a domain whose boundary consists of $g+1$ piecewise smooth closed curves. Then there exists a Riemann surface Σ of genus g with a real M-structure, and a conformal mapping $f : \Sigma^+ \to D$ that realizes a homeomorphism between the closures of Σ^+ and D.*

Let us call the map f a uniformization map for D.

PROOF. The domain D is a conformal image of a domain E bounded by circles. Let us represent the latter as a hemishell of a Riemann surface. Denote by $\tau_1, \dots, \tau_{g+1}$ the inversions of the plane with respect to the boundary circles. Let Ω be the set of all the points on the plane that hit E after undergoing a finite sequence of inversions τ_j. The group T generated by τ_j, $1 \le j \le g+1$, acts on Ω. Let $T_0 \subset T$ be the subgroup of all sense-preserving elements in T. It is easy to check that T_0 acts freely on Ω, and that for any $i \in \{1, \dots, g+1\}$, $E \cup \tau_i E$ is a fundamental domain for T_0. Let $\Sigma = \Omega/T_0$. Then Σ is a compact regular Riemann surface of genus g obtained from $E \cup \tau_i E$ by identifying the components of its boundary that are mapped into each other by the inversion τ_i. The map τ_i induces an antiholomorphic involution on Σ whose set of fixed points divides the surface into two hemishells Σ^+ and Σ^- that are permuted by this involution.

5.5. Singularity correspondence. Let us introduce the Cauchy transform of a multiply connected domain:

$$h_D(w) = \frac{1}{\pi}\int_D \frac{dxdy}{w-z}, \quad z = x+iy, \quad w \notin D. \tag{5.2}$$

The Cauchy transform has properties 1 and 3–7 from Chapter 3, analogously to the case of a simply connected domain, and consists of $g+1$ holomorphic functions defined in the connected components of the complement of D. In general, there is no relation between these functions.

Analogously to the simply connected case, a domain is algebraic if and only if its Cauchy transform is rational. Also, a domain is abelian if and only if the derivative of the Cauchy transform is rational. More precisely, the following theorem is true.

SINGULARITY CORRESPONDENCE THEOREM(Gustaffson [6]). *Let f be the uniformization mapping for a domain D that maps a hemishell Σ^+ of a Riemann surface Σ with a real M-structure onto D. Then*

1. *The function $\psi(\zeta) = \overline{f(\tau(\zeta))} - h_D(f(\zeta))$ continues to a holomorphic function in Σ^+ (here τ is an antiholomorphic involution on Σ).*
2. *The function f extends to a meromorphic function on the entire surface if and only if the Cauchy transform h_D extends to a rational function on the Riemann sphere.*
3. *The form df extends to a meromorphic differential on the entire surface if and only if the derivative of the Cauchy transform h_D extends to a rational function on the Riemann sphere.*
4. *If D is an algebraic domain then the functions f and h_D have the same degree. If D is an abelian domain then the differential forms df and dh_D have the same degree.*

PROOF. 1. Plemelj formula implies that the function $\psi(\zeta) = \overline{f(\zeta)} - h_D(f(\zeta))$ defined on the boundary of the hemishell Σ^+ extends analytically, without singularities, inside this hemishell. The extension of $\overline{f(\zeta)}$ inside the hemishell can be written as $\overline{f(\tau(\zeta))}$, since the boundary of the hemishell is fixed by τ.

2. If f is meromorphic on the entire surface then $\overline{f(\tau(\zeta))}$ is meromorphic in Σ^+, hence, $h_D(f(\zeta))$ extends to a meromorphic function in Σ^+, by statement 1 of the theorem. Therefore, h_D extends to a meromorphic function in D. Since h_D is analytic outside D, h_D has to be a rational function. Conversely, if h_D is rational then $h_D(f(\zeta))$ is meromorphic in Σ^+, which implies that $\overline{f(\tau(\zeta))}$ is a meromorphic function in Σ^+ as well, and $f(\zeta)$ is meromorphic outside Σ^+. Since we know that the latter function is analytic

inside Σ^+, we can conclude that it extends to a meromorphic function on the entire surface.

3. The proof is analogous to that of statement 2.

4. Every singular point of f on the surface Σ matches a singular point of the same type of the Cauchy transform h_D, and vice versa. Consequently, the functions f and h_D and the forms df and dh_D have the same degree.

5.6. Proof of the theorem on algebraic solutions. Theorem 5.2 implies that the Cauchy transform of the fluid domain changes in time as follows:

$$h_{D(t)}(w) = h_{D(0)}(w) + \sum_{j=1}^{n} \frac{1}{\pi(w - z_j)} \int_0^t q_j(\tau)d\tau.$$

Hence, the functions $h_{D(t)}$ and $dh_{D(t)}/dw$ are rational if and only if they are rational at $t = 0$, and $\deg h_{D(t)} \leq \deg h_{D(0)} + n$, $\deg dh_{D(t)}/dw \leq \deg dh_{D(0)}/dw + 2n$. By the singularity correspondense theorem, this implies that the uniformization map f_t and its differential df_t are meromorphic on the surface for all t if and only if they have this property at $t = 0$, and that

$$\deg f_t \leq \deg f_0 + n, \quad \deg df_t \leq \deg df_0 + 2n.$$

Q.E.D.

5.7. Construction of solutions. The above theorems on algebraic solutions and the correspondence of singularities reduce the problem of describing the dynamics of an abelian domain to the problem of solving a system of equations on the parameters of an abelian domain of a fixed multiplicity.

THEOREM (on local uniqueness). *Let $D(s)$, $s \in (s_1, s_2)$, be a smooth family of domains, $s_0 \in (s_1, s_2)$, and let $\zeta_1, \dots, \zeta_g$ be fixed points in the g distinct bounded connected components of the complement to $D(s)$. If the Cauchy transform of $D(s)$ and integrals $\int_{D(s)} \log|z - z_m|\,dx\,dy$, $m = 1, \dots, g$, are independent of s then $D(s) = \text{const}$.*

IDEA OF THE PROOF. Any harmonic function u in a neighborhood of $D(s)$ can be uniformly approximated by linear combinations of functions $\operatorname{Re}\frac{1}{w-z}$, $\operatorname{Im}\frac{1}{w-z}$, $w \notin D(s)$, and $\log|\zeta - \zeta_m|$, therefore $\int_{D(s)} u\,dx\,dy$ is independent of s. Therefore, by the local uniqueness theorem for a domain with prescribed moments, $D(s)$ has to be the same domain for all s.

According to this theorem, the parameters of the uniformization map for $D(t)$ can be found from a system of nonlinear equations that includes:

1. The conditions of cancellation of principal parts of functions $\overline{f(\tau(\zeta))}$ and $h_D(f(\zeta))$ at singular points.

2. Evolution of the moments of $D(t)$ with respect to the functions $\log|\zeta - \zeta_m|$, where ζ_m belongs to the m-th connected component of the complement to $D(t)$:

$$(5.3)\quad \int_{D(t)} \log|\zeta - \zeta_m| = \int_{D(0)} \log|\zeta - \zeta_m| + \sum_{j=1}^{n} q_j \log|z_j - \zeta_m| - 2\pi(\Phi_m - \Phi_{g+1}).$$

5.8. Reconstruction of an annular domain from its moments. Consider an algebraic domain with one hole. It is the image of a rectangle under a conformal mapping expressed by an elliptic function. Let us consider only the case when all the singularities of the Cauchy transform are simple poles.

Let us denote by $\zeta_\lambda(w)$ the Weierstrass elliptic ζ-function with periods 1 and $i\lambda$, $\lambda > 0$, which is a meromorphic function on the entire complex plane that has simple poles with residue 1 at points $m + ni\lambda$ (m, n are integers) and satisfies the periodicity conditions $\zeta_\lambda(w+1) = \zeta_\lambda(w)+\delta$, $\zeta_\lambda(w+i\lambda) = \zeta_\lambda(w) + \delta'$, where δ and δ' depend on λ ([**19**]).

THEOREM. *Let D be an algebraic domain that is homeomorphic to an annulus and has the Cauchy transform*

$$h_D(z) = \sum_{j=1}^{d} \frac{A_j}{z - B_j}. \tag{5.4}$$

Then this domain is the image of the rectangle

$$\{(x, y) \mid 0 \leq x \leq \frac{1}{2}, 0 \leq y \leq \lambda\}, \quad \lambda > 0,$$

under the conformal map

$$f(w) = \sum_{j=1}^{d} c_j \zeta_\lambda(w - \alpha_j) + C. \tag{5.5}$$

The unknown parameters c_j, α_j, C, λ can be found from the following equations:

$$f(-\overline{\alpha_j}) = B_j, \quad j = 1, \dots, d; \tag{5.6}$$

$$-\overline{c_j} = \frac{A_j}{f'(-\overline{\alpha_j})}, \quad j = 1, \dots, d; \tag{5.7}$$

$$\sum_{j=1}^{d} c_j = 0. \tag{5.8}$$

PROOF. The Riemann surface with M-structure corresponding to the domain D is a torus. One can represent this torus as a quotient space $\mathbf{C}/\Gamma$ of the complex plane by a rectangular lattice with periods 1 and $i\lambda$, and the hemishell Σ^+ corresponds to the rectangle $\{(x, y) \mid 0 \leq x \leq \frac{1}{2}, 0 \leq y \leq \lambda\}$. By the singularity correspondence theorem, the uniformization map for the domain D is expressed by a meromorphic function on Σ with d simple poles, i.e. by an elliptic function of the form (5.5), and the poles of the functions $\overline{f(\tau(\zeta))}$ and $h_D(f(\zeta))$ inside Σ^+ and their residues must coincide. The coincidence of the poles is expressed by equations (5.6), the coincidence of the residues, by equations (5.7). Equation (5.8) is the condition that the sum of all residues of an elliptic function equals zero.

In order to define uniquely the dynamics of an annular domain under injection of fluid, to equations (5.6)–(5.8) linking the uniformization map and the Cauchy transform one should add equation (5.3). In case of an annular domain it has the form

$$\int_0^{\frac{1}{2}}\int_0^{\lambda} |f_t'(w)|^2 \log|f_t(w)|\,dx\,dy = \int_0^{\frac{1}{2}}\int_0^{\lambda} |f_0'(w)|^2 \log|f_0(w)|\,dx\,dy + \sum_{j=1}^{n} q_j \log|z_j| - 2\pi(\Phi_1 - \Phi_2). \tag{5.9}$$

By Theorem 5.7, equations (5.6)–(5.9) define the domain $D(t)$ in a locally unique way.

EXAMPLE. Consider an annular domain D with the Cauchy transform

$$h_D(w) = \frac{A}{\pi}\sum_{k=0}^{N-1} \frac{1}{w - Re^{2\pi ik/N}}, \tag{5.10}$$

and such that $\int_D \log \mid \frac{z}{R} \mid dxdy = 0$ (Figure 16). Such a domain exists on some interval of values of the ratio A/R^2, and it is invariant with respect to the rotation through the angle $2\pi/N$ about the origin. Consider the evolution of this domain as the fluid is injected from N sources situated at points $Re^{2\pi ik/N}$, i.e. at vertices of a regular N-gon, at the same rate q, assuming the velocity potential on both components of the boundary to be zero. The Cauchy transform of $D(t)$ is

$$h_{D(t)}(w) = \frac{A+qt}{\pi}\sum_{k=0}^{N-1} \frac{1}{w - Re^{2\pi ik/N}}.$$

By formula (5.5), the uniformization map has to be

$$f_t(w) = \sum_{k=0}^{N-1} Be^{2\pi ik/N}\zeta_\lambda\left(w + \alpha + \frac{ik\lambda}{N}\right), \quad B, \alpha, \lambda \in \mathbf{R}.$$

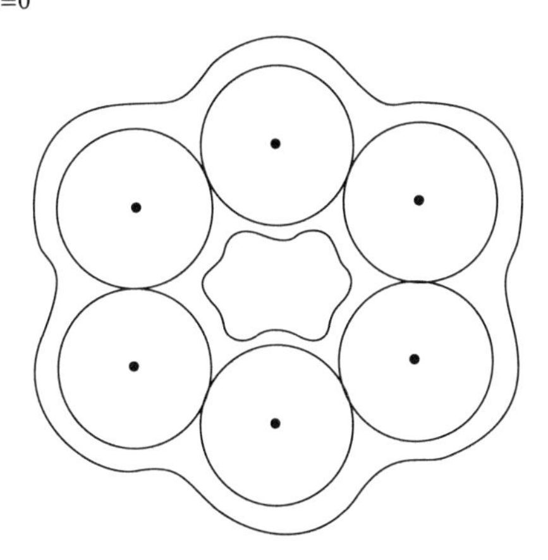

FIGURE 16

The unknown parameters B, α, λ have to be determined from equations (5.6)–(5.9):

$$f_t(\alpha) = R, \quad f_t'(\alpha) = -\frac{qt + A}{\pi B};$$

$$\int_0^{\frac{1}{2}} \int_0^{\lambda} |f_t'(w)|^2 \log |f_t(w)/R| dx dy = 0.$$

Problems. 1. If a domain is bounded by circles and is not simply connected, it is not abelian. For example, a circular annulus is not an abelian domain.

2. The boundary of an algebraic domain is an algebraic curve (possibly disconnected).

3. (Gustafsson). Any algebraic domain of degree 2 is simply connected.

4. A domain is algebraic if and only if the integral over this domain of any holomorphic function is a linear combination of its derivatives at a finite number of points with coefficients independent of the choice of the function.

5. For a sufficiently large n, there exists an algebraic domain of genus g whose Cauchy transform is a polynomial of degree n in $1/z$.

6. Any finitely connected domain with a smooth boundary can be (uniformly) approximated by algebraic ones as precisely as needed.

7. (P. P. Kufarev). Let the fluid initially saturate the strip $|\operatorname{Im} z| < b$, and let the injection points be $z = na$, where n runs through all integers. The rate of injection is q. Find the fluid domain at a time t.

8. Consider the evolution of an annular domain bounded by two circles produced by extraction of gas from the hole. Find the contraction point of the inner component of the boundary (Figure 17).

9. Let D be a doubly connected domain that does not contain the origin, and assume that $\int_D z^n dx dy = 0$ for all nonzero integers n. Then D is a circular annulus centered at the origin.

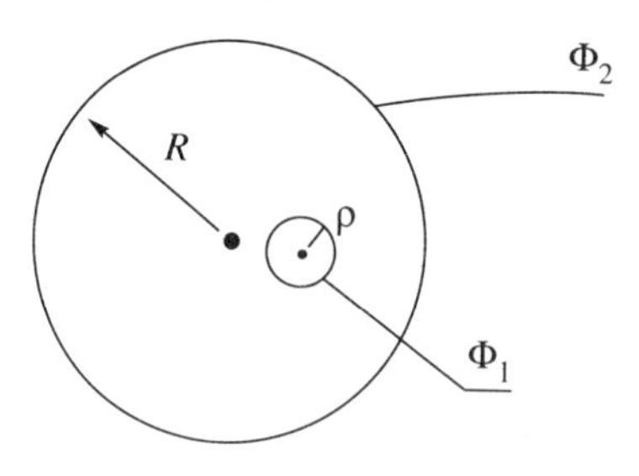

FIGURE 17

6. Evolution with Topological Transformations

6.1. Weak solutions of the evolution problem. In the problem of injection into a multiply connected domain it is natural to assume that gas is extracted from all holes under the same pressure. For instance, in the problem of filling a mould (see §7.1) the pressure on all components of the boundary is athmospheric.

Thus, consider a special case of evolution of a multiply connected domain, namely, when the rates of the sources are positive and the velocity potential equals zero on all components of the boundary. In this case, the solution $D(t)$ of the problem possesses the"inclusion property" (compare to §4.2):

THEOREM. $D(t_1) \subset D(t_2)$ *if* $t_1 < t_2$.

In the process of injection the fluid domain may transform topologically. For example, parts of the boundary may collide (Figure 18). In order to describe mathematically the evolution after the time of collision, let us introduce the notion of a weak solution. Solutions that we have considered so far will be called classical.

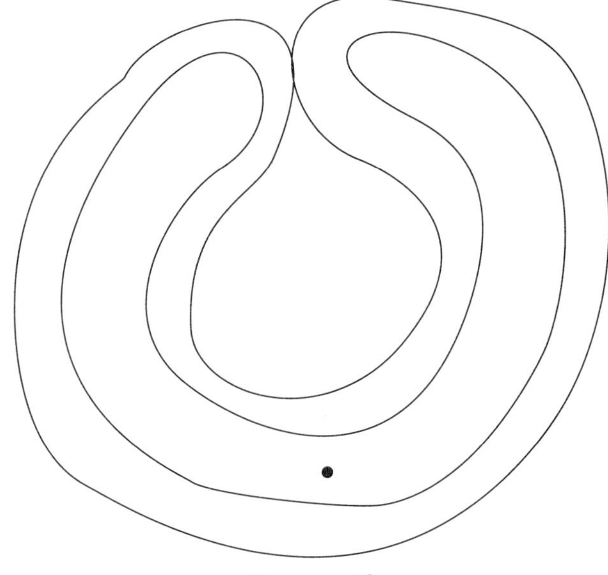

FIGURE 18

DEFINITION. Let us call a family of bounded domains $D(t)$, $0 \le t < \infty$, a *weak solution* of the injection problem if it has the following properties:

1) inclusion: $D(t_1) \subset D(t_2)$ if $t_1 < t_2$;

2) linearly growing area: the area of $D(t)$ equals $S_0 + (\sum_{j=1}^n q_j)t$, S_0 being the area of $D(0)$;

3) finite number of topological transformations: there exist times τ_1, ..., τ_k such that on each of the $k+1$ time intervals between them: $(0, \tau_1)$, (τ_1, τ_2), ..., (τ_{k-1}, τ_k), (τ_k, ∞), the domain $D(t)$ is bounded by a finite collection of disjoint smooth nonsingular curves;

4) piecewise classical solution: on each of the $k + 1$ time intervals, the family of domains $D(t)$ is a classical solution of the injection problem.

REMARK. If t_1 and t_2 belong to different intervals, the domains $D(t_1)$ and $D(t_2)$ may, in general, have different number of components and holes.

The theorem on first integrals is true for weak solutions.

THEOREM. *Let $u(x, y)$ be a harmonic function in a domain that contains $D(t)$ together with its boundary. Then*

$$\frac{d}{dt}\int_{D(t)} u\,dx\,dy = \sum_{j=1}^{n} q_j\,u(z_j). \tag{6.1}$$

Indeed, between the singular times τ_j this equality holds by Theorem 5.2; hence it is true for all t because the moments depend on time continuously.

COROLLARY.

$$\int_{D(t)} u\,dx\,dy = \int_{D(0)} u\,dx\,dy + \sum_{j=1}^{n} u(z_j)\int_0^t q_j(\tau)\,d\tau. \tag{6.2}$$

In order to describe the evolution of a fluid domain, we need to be able to construct the weak solution corresponding to a given initial domain and prescribed positions and rates of sources. It has been proved in [**13**] that this solution exists and is unique assuming that the initial domain is algebraic. It seems very natural that this should be true for any finitely connected domain with a smooth boundary, though we are not aware of any rigorous proof of this statement.

For $t > \tau_k$, the domain $D(t)$ is simply connected, and for large t it has an approximately round shape.

Formula (6.2) implies that the result of injection depends only on the total amounts of fluid injected from each of the sources, not on the order or schedule of their work, just as in the classical case.

REMARK. In case when the pressure takes different values on different components of the boundary, it does not seem possible to define a weak solution with satisfactory properties.

At times τ_j the fluid domain undergoes some topological transformations.

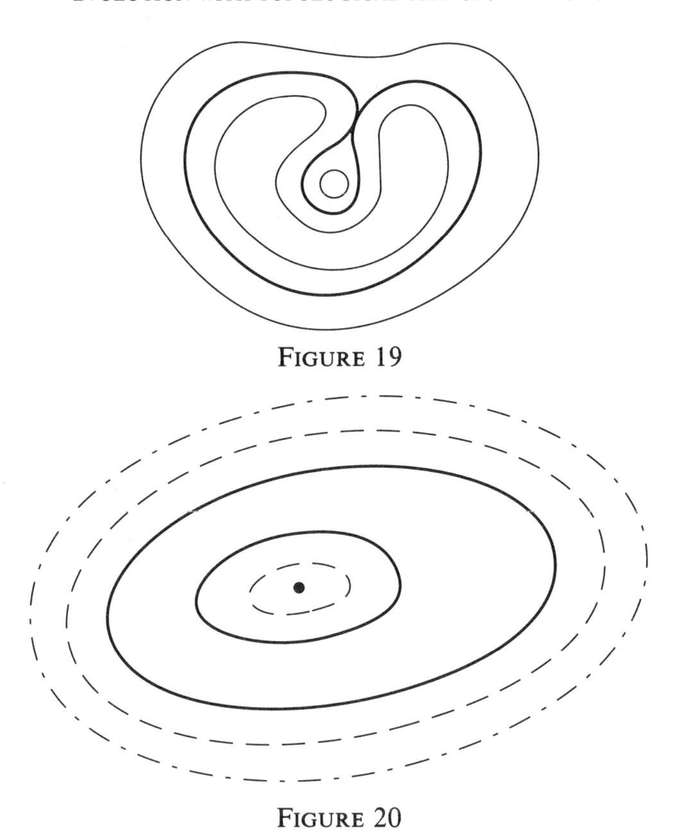

FIGURE 19

FIGURE 20

There are two kinds of such transformations:

1. Collision of two parts of the boundary (Figure 19).
2. Contraction of a hole (Figure 20).

A hole always contracts to a single point (cf. §4.4).

Transformations of the first and second kind may be interlaced which will result in changing the number of connected components of the domain and its boundary in the course of evolution.

The topological structure of a simplest weak solution is pictured in Figure 21.

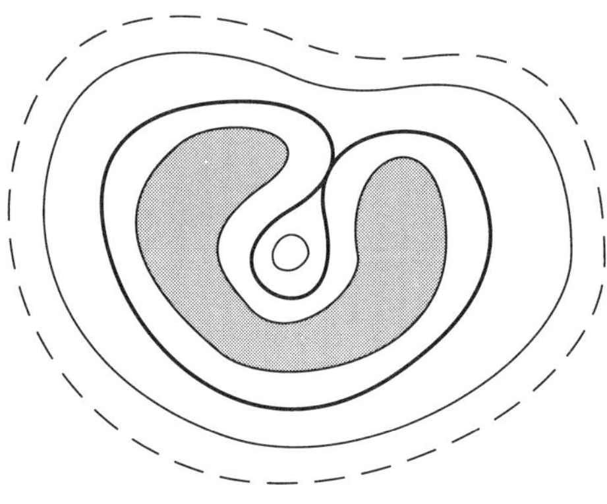

FIGURE 21

6.2. The simplest algebraic solutions. Let $D(t)$ be a weak solution of the injection problem. By Theorem 6.1, the Cauchy transform of the domain $D(t)$ changes as follows:

$$h_{D(t)}(w) = h_{D(0)}(w) + \sum_{j=1}^{n} \frac{1}{\pi(w - z_j)} \int_0^t q_j(\tau)d\tau, \quad w \in \overline{D(t)}.$$

Therefore, the functions $h_{D(t)}$ and $\frac{dh_{D(t)}}{dw}$ are rational if and only if they are rational at $t = 0$. Hence, by the singularity correspondence theorem, the properties of a domain to be algebraic and abelian are preserved in the course of evolution:

THEOREM. *Let $D(t)$ be a weak solution of the evolution problem. Then*:

1. If $D(0)$ is an algebraic domain of degree d then $D(t)$ is an algebraic domain of a degree no higher than $d + n$.
2. If $D(0)$ is an abelian domain of multiplicity k then $D(t)$ is an abelian domain of a multiplicity no higher than $k + 2n$.

Consider examples of weak solutions.

EXAMPLE 1 (S. Richardson). Consider the problem of injection from two symmetric sources situated at points a and $-a$ at the same rate q. Suppose that the initial domain is an empty set. Hence, for $t \le \pi a^2/q$, the fluid domain will be the union of two disks of equal radius centered at a and $-a$. At the time $t_0 = \pi a^2/q$, the disks will unite and transform into a connected fluid domain (Figure 22). Consider further evolution of this domain, $D(t)$. The Cauchy transform of $D(t)$ equals

$$h_{D(t)}(w) = \frac{qt}{\pi}\left(\frac{1}{w-a} + \frac{1}{w+a}\right).$$

Hence, the uniformization mapping has two simple poles, so it can be sought in the form (due to the symmetry):

$$f_t(\zeta) = \frac{A_t\zeta}{1-\alpha_t\zeta} + \frac{A_t\zeta}{1+\alpha_t\zeta} = \frac{2A_t\zeta}{1-\alpha_t^2\zeta^2}, \qquad A_t > 0, \quad \alpha_t > 0.$$

The coefficients A_t and α_t can be found from equations (3.5), (3.6):

$$\frac{2A_t\alpha_t}{1-\alpha_t^4} = a, \quad 2A_t^2\frac{1+\alpha_t^4}{(1-\alpha_t^4)^2} = \frac{qt}{\pi},$$

which yield

$$A_t = \frac{a}{2}\frac{1-\alpha_t^4}{\alpha_t}, \quad \alpha_t = (T - (T^2-1)^{1/2})^{1/2}, \quad T = \frac{qt}{\pi a^2}.$$

REMARKS. 1. The problem considered is equivalent to the problem of injection from one source into a half-plane with an impervious boundary.

2. The time when the fluid domain becomes convex is $3\pi a^2/q$; at this time the width of the domain is twice the distance between the sources (Figure 22).

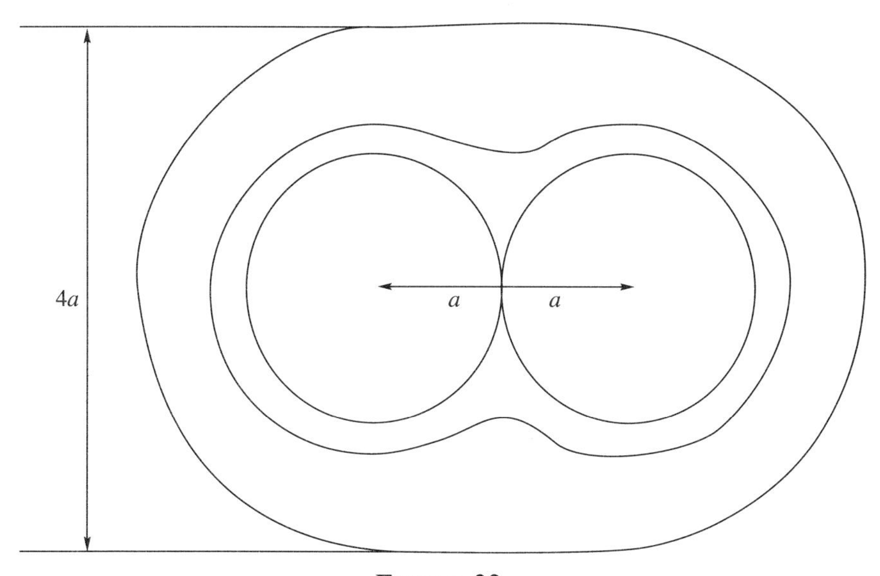

FIGURE 22

EXAMPLE 2 (Richardson). Consider the problem of injection of fluid at the same rate q from four sources situated at the four vertices of a rectangle. Let the coordinates of the sources equal $(\pm a, \pm b)$, $a \geq b$ (Figures 23, p. 48, and 24). According to Remark 2 from the previous example, if $a \geq 2b$, then first there will be four round fluid blobs; then they will unite into two pairs at the time $t_0 = \pi b^2/q$, forming two identical domains whose evolution is described in Example 1 and finally, at a time $t_1 \geq 3\pi a^2/q$ (which can be easily found from the formulas of Example 1) these two domains will unite into a simply connected domain. Let us study its dynamics. The Cauchy

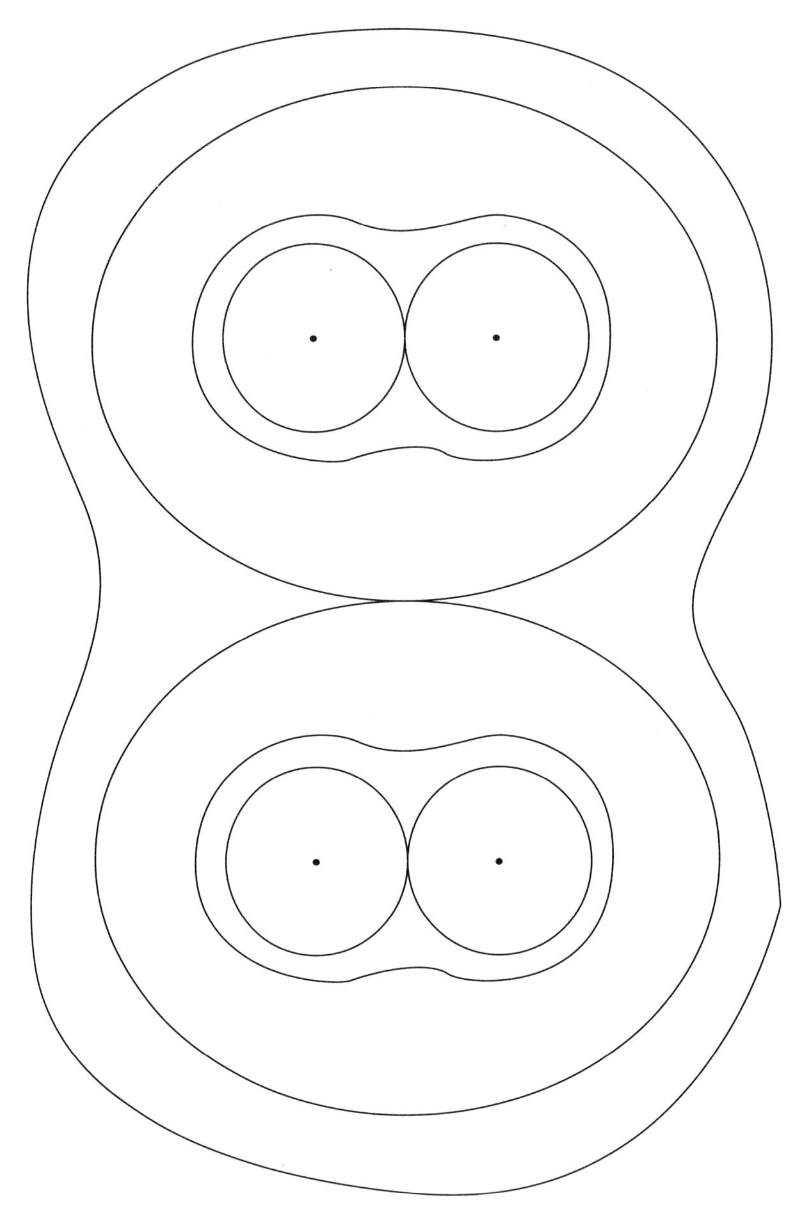

FIGURE 23

transform equals

$$h_{D(t)}(w) = \frac{qt}{\pi}\left(\frac{1}{w-a-ib} + \frac{1}{w+a-ib} + \frac{1}{w-a+ib} + \frac{1}{w+a+ib}\right).$$

Hence, the uniformization map f_t has four simple poles, so because of the symmetry

$$f_t(\zeta) = \frac{A_t\zeta}{1-(\alpha_t+i\beta_t)\zeta} + \frac{A_t\zeta}{1+(\alpha_t+i\beta_t)\zeta} + \frac{A_t\zeta}{1-(\alpha_t-i\beta_t)\zeta} + \frac{A_t\zeta}{1+(\alpha_t-i\beta_t)\zeta},$$

where A_t, α_t, β_t can be found from the equations

$$f_t(\alpha_t + i\beta_t) = a + ib, \qquad \frac{qt}{\pi} = A_t f'(\alpha_t + i\beta_t).$$

If $a < 2b$ then the second confluence happens before the critical time $3\pi a^2/q$ when each of the two uniting blobs would have become convex if they were not to collide. Therefore, the two boundaries collide at two points simultaneously, which results in formation of a doubly connected domain (Figure 24). Obviously, its Cauchy transform is the same as in case of confluence into a simply connected domain. Consider, for instance, the case $a = b$. Then the Cauchy transform equals (5.10) for $N = 4$, and the solution coincides with the dynamics of the domain in the example of §5.7, up to the time t_2 of contraction of the inner component of the boundary. After this time, the fluid domain is connected and simply connected, and the above formulas define a conformal map of the disk onto this domain. The time t_2 can be found from equations (5.6)–(5.9), in which λ tends to infinity (because $\lambda = \infty$ corresponds to the time of contraction).

REMARK. The problem we just considered is equivalent to the problem of injection from one source into a right angle with impervious boundaries.

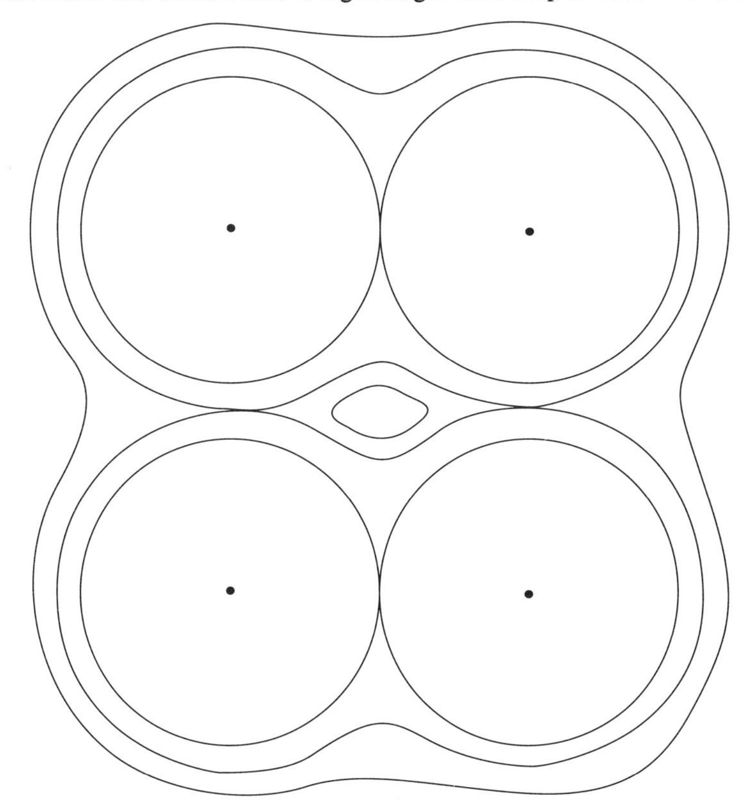

FIGURE 24

6.3. Weak solutions of the contraction problem. In the problem of contraction considered in Chapter 4, let us study the dynamics of gas domains that do not remain connected in the course of evolution (see §4.2). That means, at some point, two parts of the boundary collide (Figure 25). Define the dynamics of the domain after the time of collision, up to the time t^* when gas will be extracted entirely. We will do this by introducing the concept of a weak solution analogously to §6.1.

DEFINITION. Let us say that a family of bounded domains $D(t)$, $0 \leq t < t^*$, consisting of a finite number of connected components is a *weak solution* of the contraction problem if it has the following properties:

1) inclusion: $D(t_2) \subset D(t_1)$ if $t_1 < t_2$;

2) linearly decreasing area: the area of $D(t)$ equals $S_0 - qt$, where S_0 is the area of $D(0)$, and q is the rate of extraction;

3) finite number of topological transformations: there exist times $\tau_1, \dots, \tau_k$ such that on each of the $k+1$ time intervals between them: $(0, \tau_1)$, (τ_1, τ_2), $\dots$, (τ_{k-1}, τ_k), (τ_k, ∞), the domain $D(t)$ is bounded by a finite collection of disjoint smooth nonsingular curves;

4) piecewise classical solution: on each of the $k+1$ time intervals, the family of domains $D(t)$ is a classical solution of the contraction problem (taking into account the remark in §4.1).

A weak solution $D(t)$ corresponding to a given initial domain $D(0)$ exists and is unique if $D(0)$ is algebraic. It is natural to believe that the same is true for an arbitrary smooth initial domain, but a rigorous proof has not been worked out.

Let us say that a given point is a point of full contraction if it belongs to $D(t)$ for all t from 0 to t^*, t^* being the time of contraction. There can be several such points since the gas domain can break up into several parts.

MAIN THEOREM. *Let $D(t)$ be a weak solution of the contraction problem. Then*:

1. *The gravity potential $\Pi_{D(t)}$ of the domain inside itself that is given by formula* (2.3) *changes by a constant in the course of contraction*: $\Pi_{D(0)} - \Pi_{D(t)} = \text{const}(t)$ in $D(t)$.

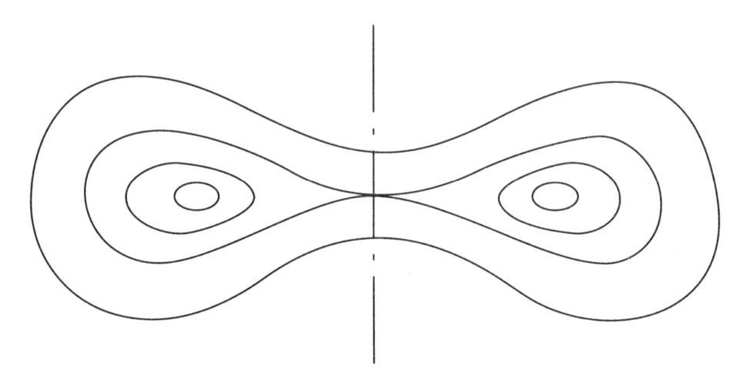

FIGURE 25

2. The set of points of full contraction is finite.

3. The minimal value of the gravity potential of a domain is attained at the points of full contraction, and only there.

This theorem is proved analogously to Theorem 4.4.

REMARK. Existence of several points of full contraction is a sufficient condition for the domain to break up in the course of contraction. This condition can be effectively verified since the points of full contraction are the global minima of the potential. It is especially useful for domains that have some kind of geometric symmetry, because in the class of such domains, several global minima may appear in general position.

A weak solution of the contraction problem with sources is defined analogously. In this case, the set of points of full contraction coincides with the set of global minima of the "effective potential" whose relationship with the gravity potential of the initial domain is given by formula (4.16).

6.4. A sufficient condition of a breakup of a symmetric domain. Let the boundary of the initial gas domain have the equation $y^2 = f(x)$, where $f(x)$ is an even function (Figure 26).

THEOREM. *If*

$$\int_0^b \frac{xf'(x) + 2f(x)}{f(x)^{1/2}(x^2 + f(x))} dx > \frac{\pi}{2},$$

where b *is the smallest positive root of* f, *then the domain will break up into two pieces in the course of evolution.*

PROOF. Because of the symmetry, if the domain does not break up, it has to contract completely at the origin. The main theorem tells us that the origin then has to be a global minimum point of the gravity potential. Hence, the

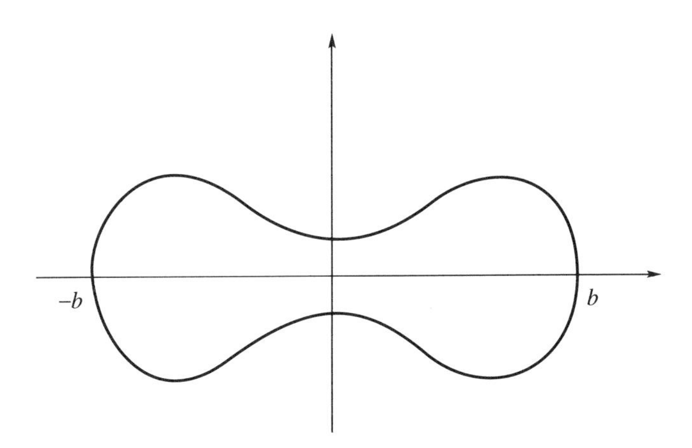

FIGURE 26

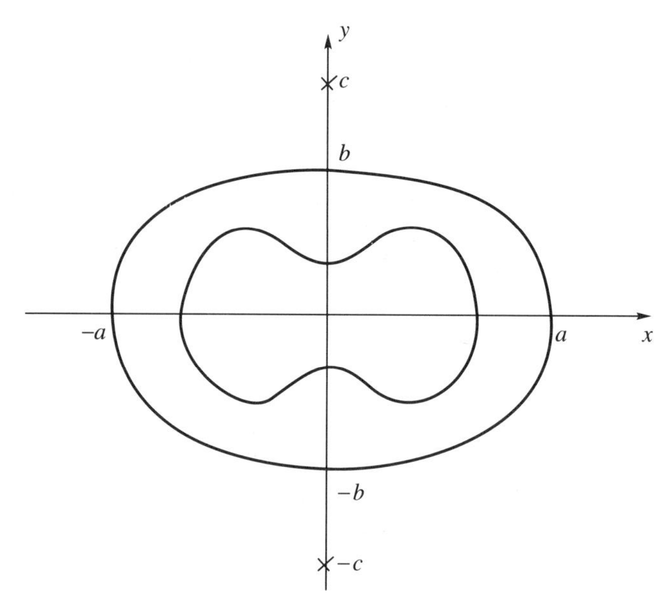

FIGURE 27

second derivative $\frac{\partial^2 \Pi_D}{\partial x^2}(0, 0)$ has to be nonnegative. A calculation shows that

$$\frac{\partial^2 \Pi_D}{\partial x^2}(0, 0) = 1 - \frac{2}{\pi} \int_0^b \frac{x f'(x) + 2f(x)}{f(x)^{1/2}(x^2 + f(x))} dx,$$

which proves the theorem.

EXAMPLE. Let the initial gas domain be an ellipse with half-axes a and b, and let two sources be situated at symmetric points of the line containing the shorter axis, at a distance c from the center (Figure 27). The effective potential in this case, in a suitable coordinate system, equals

$$\widehat{\Pi} = \frac{1}{2}\left(\frac{a}{a+b}x^2 + \frac{b}{a+b}y^2\right) - \frac{ab}{8}(\log(x^2 + (y-c)^2) + \log(x^2 + (y+c)^2)).$$

In order to find out if the domain will break up, we should consider the value of $\frac{\partial^2 \widehat{\Pi}_D}{\partial x^2}(0, 0)$, and compare it to zero. The sufficient condition of a breakup then comes in the form $b(a+b) > 2c^2$. For example, if $b = a/2$ then a breakup necessarily occurs for $c < \sqrt{\frac{3}{2}}b \approx 1.22\,b$. One can also show that otherwise a breakup does not occur.

Problems. 1. Suppose that fluid is injected into an annular domain $R_1 \leq |z| \leq R_2$ from a single source working at a rate q. Show that if t is large enough, the fluid domain at the time t coincides, up to a translation, with the domain $D(t - \pi R_1^2/q)$ of Example 1 in §3.7. (Note that an analytic representation of the complete solution to this problem is unknown.)

2. Write down the equation of the boundary for the domain of Example 1 in §6.2.

3. Give an example of an algebraic domain with g holes of degree not higher than $[(g+5)/2]$ ($[a]$ stands for the integer part of a).

4. Find the points of full contraction for the domain of the example in §6.4.

5. Will the domain shown in Figure 28 break up in the course of contraction? (If worried about smoothness, make it smooth by a small perturbation.)

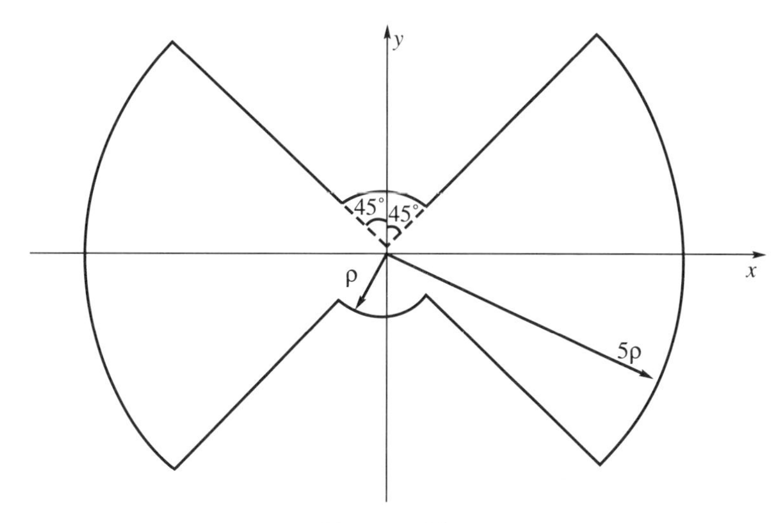

FIGURE 28

7. Contraction Problem on Surfaces

7.1. Physical motivation. The problem of evolution of a fluid domain on a surface arises in the study of the process of filling a mould shaped as a very narrow slot between two almost parallel surfaces, with molten metal or polymer [7]. Let the mould be shaped after a surface Σ in $\mathbf{R}^3$. Let us suppose, for the sake of simplicity, that this surface is closed (Figure 29). Assume that the mould is being filled by means of injection of fluid from sources situated at points $A_1, \dots, A_n$, at rates $q_1, \dots, q_n$, respectively. Since initially the mould is filled with air, this air has to be taken away through an opening in the surface. It is obvious that the optimal position of this opening is at the point of contraction of the boundary between air and fluid at the time when the fluid has filled the entire mould.

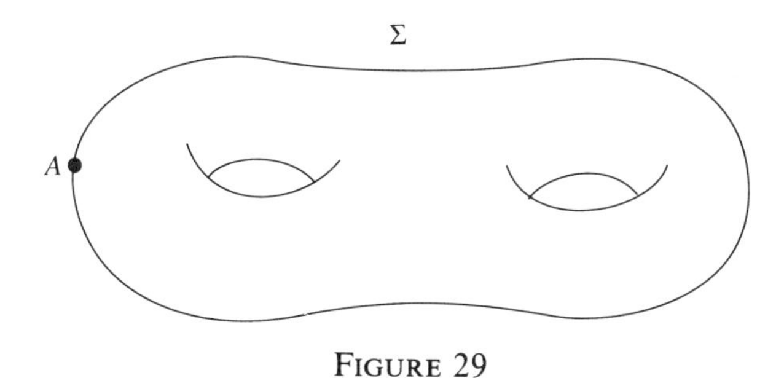

FIGURE 29

7.2. Potential and contraction points. The mathematical formulation of the problem is similar to its formulation for the "flat" case. At a time t fluid occupies a domain $D(t)$, and air occupies the domain $\Sigma\backslash D(t)$. Inside $D(t)$ the fluid velocity potential $\Phi(P, t)$ is defined satisfying the condition

$$\Delta\Phi = 0 \quad \text{in } D(t)\backslash\{A_1, \dots, A_n\}, \qquad \Phi(P, t) \sim \frac{q_j}{2\pi}\log\rho(P, A_j),\ P \to A_j,$$

q_j and A_j being the rates and the points of location of the sources, and $\rho(P, Q)$ being the length of the shortest geodesic connecting P and Q. Here Δ is Laplace's operator on the surface that acts according to the formula

$$\Delta\phi(P) = \lim_{\sigma(\Gamma)\to 0}\left(\int_\Gamma \frac{\partial\phi}{\partial n}dl/\sigma(\Gamma)\right),$$

where Γ is a small contour surrounding the point P, $\sigma(\Gamma)$ is the area that it bounds, and ϕ is a smooth function on the surface.

The boundary velocity equals $\operatorname{grad}\Phi$.

It is easy to see that any solution of this problem has the inclusion property: $D(t_1) \subset D(t_2)$ if $t_1 < t_2$. It allows to define a weak solution by analogy with §6.1. It is defined on the time interval $[0, t^*)$, where $t^* = S_a/(\sum q_j)$, and S_a is the area of the air domain.

THEOREM. *Let $u(P)$ be a smooth function on the surface defined in a neighborhood of $D(t)$ and such that $\Delta u(P) = 0$. Then*

$$\frac{d}{dt}\int_{D(t)} u(P)\,d\sigma = \sum_{j=1}^{n} q_j\, u(A_j), \tag{7.1}$$

where $d\sigma$ is an element of area on the surface.

This theorem is proved analogously to Richardson's theorem.

REMARK. For surfaces of revolution, it turns out to be possible to construct explicit solutions of the injection problem that are counterparts to the algebraic solutions of Chapter 3.

Consider the physically most interesting case of injection into an "empty" mould: the initial fluid domain $D(0)$ will be an empty set. Define the generalized gravity potential for the injection problem as a solution of the following boundary value problem for Poisson's equation:

$$\begin{gathered}\Delta\Pi = 1 \quad \text{in } \Sigma\backslash\{A_1, \dots, A_n\},\\ \Pi(P) \sim -\frac{q_j}{q_1 + \cdots + q_n}\frac{S}{2\pi}\log\rho(P, A_j), \qquad P \to A_j,\end{gathered} \tag{7.2}$$

where S is the total area of the surface.

Let us say that a point on the surface is a point of full contraction if it belongs to the air region for all times t between 0 and $t^* = S/(\sum q_j)$.

THEOREM. 1. *The generalized gravity potential inside the air region changes by an additive constant in the course of contraction.*

2. *The set of points of full contraction is finite.*

3. *Any point of full contraction is a global minimum point of the generalized gravity potential, and vice versa.*

Thus, we have reduced the problem of finding the points of full contraction to the problem of calculation of the generalized gravity potential corresponding to a system of sources.

7.3. Calculation of the generalized gravity potential on the surface. Let a surface Σ have the topology of a sphere; then by Riemann's theorem, it is conformally equivalent to the sphere, i.e. there exists a smooth, one-to-one map $z: \Sigma \to \overline{\mathbf{C}}$ that preserves angles between curves [20], [21]. The complex number $z(P)$ will be called the complex coordinate of a point P. The map z can be chosen so that $z(A) = \infty$. Let $x = \operatorname{Re} z$, $y = \operatorname{Im} z$ be the real coordinates on the surface. If the area element on the surface is $g(x, y)\, dx\, dy$ then Laplace's operator in coordinates x, y has the form

$$\Delta \phi(x, y) = \frac{1}{g(x, y)} \left(\frac{\partial^2 \phi(x, y)}{\partial x^2} + \frac{\partial^2 \phi(x, y)}{\partial y^2} \right).$$

If the only source is situated at infinity then the generalized gravity potential is defined by the conditions

$$\Delta \Pi = g(x, y), \quad x, y \in \mathbf{R}; \qquad \Pi \sim \frac{q}{2\pi} \log |z|, \quad z \to \infty. \tag{7.3}$$

If the sources are situated at points $\infty, z_1, \dots, z_n$ and have rates $q_1, \dots, q_n$ then the generalized gravity potential is defined by the conditions:

$$\begin{gathered} \Delta \widehat{\Pi} = g(x, y) \text{ outside the injection points}, \\ \widehat{\Pi} \sim -\frac{q_j}{q + q_1 + \cdots + q_n} \frac{S}{2\pi} \log |z - z_j|, \quad z \to z_j. \end{gathered} \tag{7.4}$$

It is obvious that

$$\widehat{\Pi} = \Pi - \sum_{j=1}^{n} \frac{q_j}{q + q_1 + \cdots + q_n} \frac{S}{2\pi} \log |z - z_j|. \tag{7.5}$$

Now suppose that Σ is a surface of revolution and that A is one of its poles (Figure 30). Then the complex coordinate z can be written as $z = \rho e^{i\theta}$, ρ and θ being the longitude and the latitude coordinates. The generalized gravity potential Π is invariant with respect to rotations, therefore it can be expressed in terms of ρ only. According to (7.3)

$$\frac{1}{\rho} \frac{d}{d\rho} \left(\rho \frac{d\Pi}{d\rho} \right) = g(\rho).$$

Taking into account the asymptotics at infinity, we obtain

$$\Pi(\rho) = \int_0^{\rho} \frac{1}{\xi} \int_0^{\xi} g(\eta)\, d\eta d\xi + C. \tag{7.6}$$

So, if the function $\rho(P)$ is given then the generalized gravity potential of a system of sources can be effectively calculated by (7.5) and (7.6).

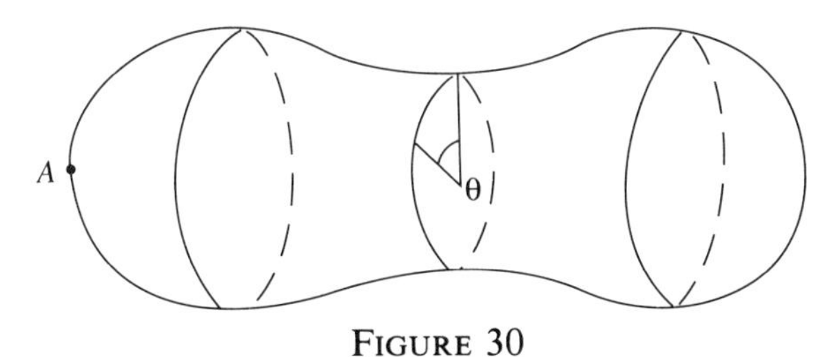

FIGURE 30

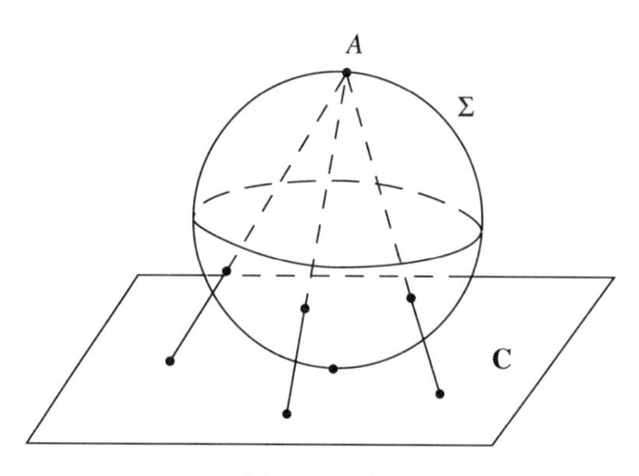

FIGURE 31

EXAMPLE. Let Σ be the sphere of a radius R. Then we can take z to be the usual stereographic coordinate (Figure 31). In this case we have

$$g(\rho) = 4R^2/(1 + \frac{\rho^2}{R^2})^2.$$

After a simple algebra we obtain an expression for the generalized gravity potential:

$$\Pi(\rho) = R^2 \log\left(1 + \frac{\rho^2}{R^2}\right) + C. \tag{7.7}$$

For the case of several sources we have

$$\begin{aligned} \widehat{\Pi}(x, y) &= R^2 \log\left(1 + \frac{x^2 + y^2}{R^2}\right) \\ &\quad - R^2 \sum_j \frac{q_j}{q + q_1 + \cdots + q_n} \log \frac{(x - x_j)^2 + (y - y_j)^2}{R^2} + C. \end{aligned} \tag{7.8}$$

If there are two sources, A and A_1, with rates q and q_1, and if the coordinate of A_1 is a real positive number a then

$$\widehat{\Pi}(x, y) = R^2 \log\left(1 + \frac{x^2 + y^2}{R^2}\right) - R^2 \frac{q_1}{q + q_1} \log \frac{(x - a)^2 + y^2}{R^2} + C.$$

Solving the equation $\frac{\partial \widehat{\Pi}}{\partial x}(x, 0) = 0$, we derive an expression for the coordinate of the contraction point:

$$x = \frac{a(q + q_1)(a^2(q + q_1)^2 + 4qq_1R^2)^{1/2}}{2q}.$$

7.4. Breakup of the boundary on a symmetric surface of revolution. Let Σ be a surface of revolution, symmetric with respect to a plane orthogonal to the axis of revolution. Such a surface can be thought of as a result of rotating a graph of an even function $\phi(x)$ about the abscissa axis (Figure 32). The section of our surface by the symmetry plane will be called the equator. Suppose that the fluid is injected from a point A lying on the equator. If the surface is a sphere or close to a sphere, a breakup of the boundary will not occur, and it will contract to the point B antipodal to A (Figure 33). But if the surface is shaped as a very prolate ellipsoid, the picture will be different because of a breakup of the boundary in the process of contraction (Figure 34). An astonishing thing is that we can distinguish very easily between these two situations.

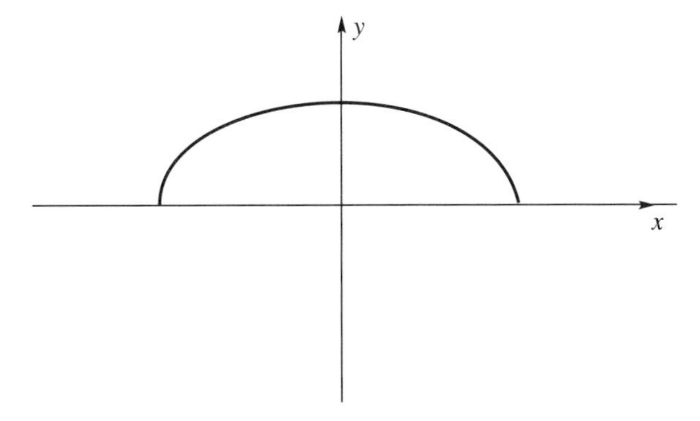

FIGURE 32

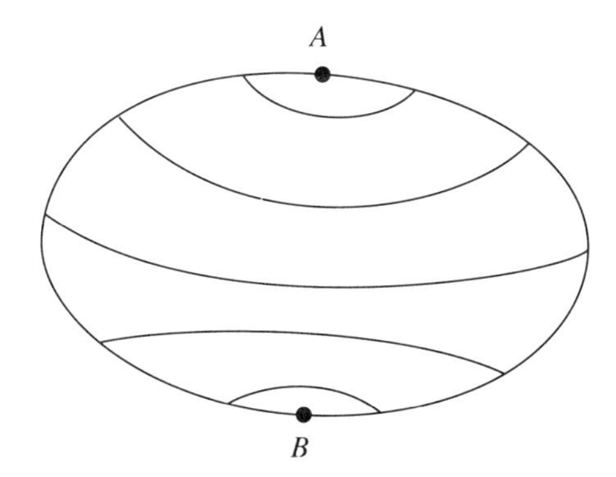

FIGURE 33

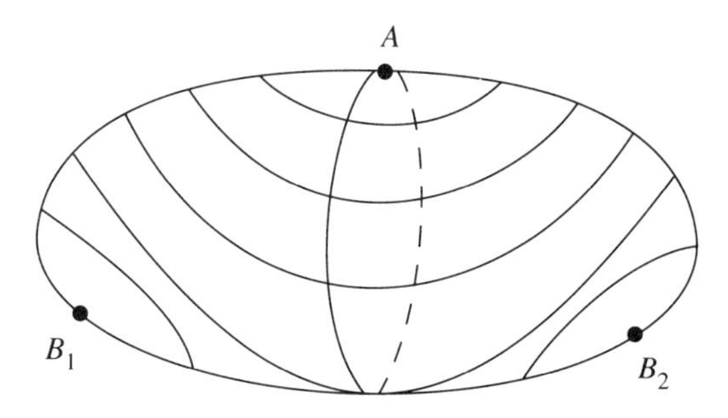

FIGURE 34

THEOREM. *If the area of the surface exceeds eight times the area of the equatorial section then the contour will break up in the course of contraction.*

PROOF. If the contour did not break up, it would contract to the point B. With the help of the formulas of §7.3 it is easy to find that this point is a local minimum point of the generalized gravity potential (i.e. $d^2\hat{\Pi}(B) \geq 0$) if and only if the ratio of areas of the surface and equatorial section is less than or equal to eight (check it!).

EXAMPLE. On an ellipsoid of revolution with half-axes a and b, the boundary breaks up in the course of contraction if (and, in fact, only if) $a/b > \lambda \approx 3.418$.

Problems. 1. A mould consists of two hemispheres and a cylinder whose height is greater than radius (Figure 35). Will the air region break up when filling such a mould from an equatorial point?

2. It can be shown that at times close to the time of full contraction, the boundary of the air region has an approximately elliptic shape (cf. §4.10). Find the ratio of the axes of the asymptotic ellipse for the problem of filling a spherical mould from two sources working at equal rates (Example 7.3, $q = q_1$).

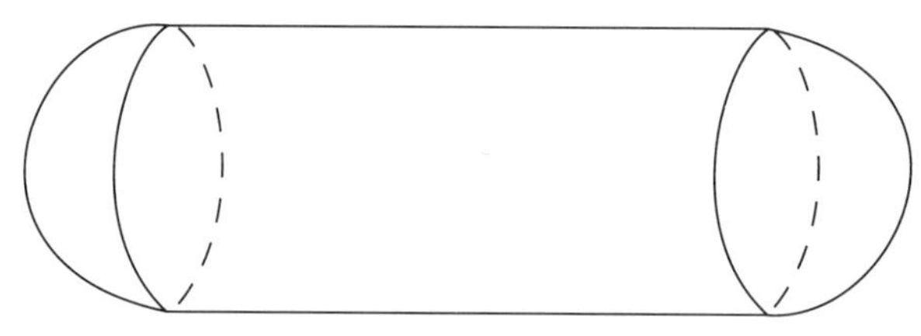

FIGURE 35

Answers and Clues to the Problems

Chapter 2. 1. *Answer*: $z(t) = \frac{Sz_0 - aqt}{S-qt}$, where a is the coordinate of the sink, q is its rate, t is time, S is the area of the disk, z_0 is the coordinate of its center.

2. The coordinates of the mass center of the domain can be expressed in terms of its moments. The mass center of a convex domain lies inside it. Hence, the two domains considered must have a common point (the mass center), with respect to which they are starlike, therefore by the theorem of Novikov, they have to coincide.

3. Take three domains D_1, D_2, D_3 situated as shown in Figure 36, and containing a common point Q. Let us inject in them a certain (the same for all three domains) amount of fluid at this point. We will obtain new

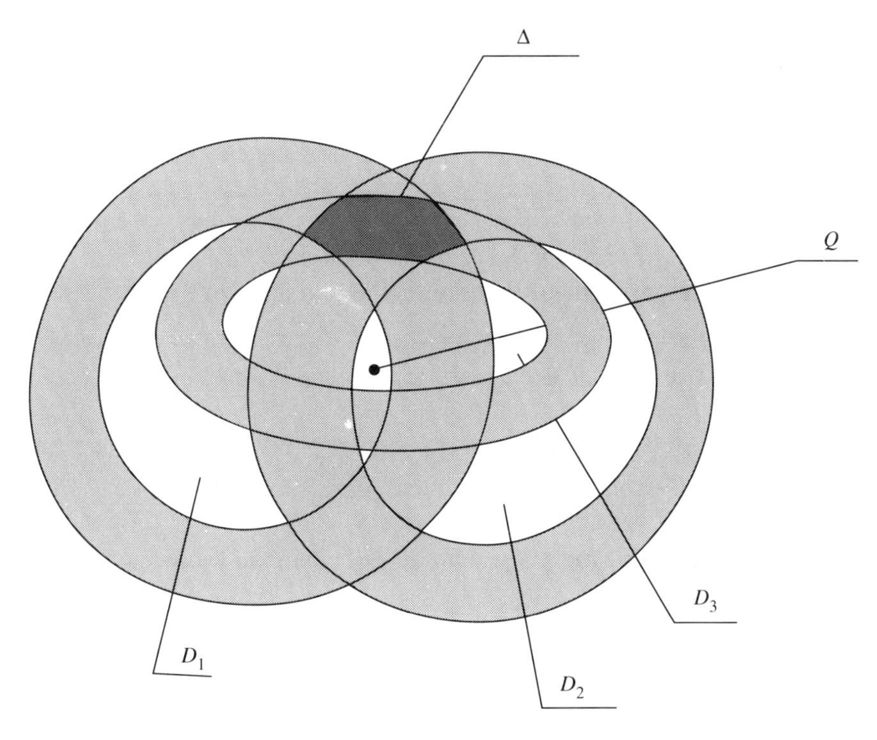

FIGURE 36

domains D_1', D_2', D_3'. Consider the annular domains $D_1'\backslash D_1, D_2'\backslash D_2, D_3'\backslash D_3$. They have the same moments. The intersection of these domains is a hexagon Δ. Eliminating this hexagon from the three annular domains, we will obtain three simply connected domains with the same moments. Analogously one can construct any number of simply connected domains with equal moments and even an infinite collection of such domains. L. Zalcman, using another method, recently managed to construct an uncountable collection of this type [22]. For more details on this subject see [23].

Chapter 3. 1. ∂D is the image of a curve of order two (circle) under a rational map of degree $\deg D$, therefore the degree of ∂D must be $2\deg D$.

2. a) The boundary of an algebraic domain is a rational algebraic curve since it is the image of the circle under a rational map. But it is known that the curve $x^4 + y^4 = 1$ is not rational.

b) The Cauchy transform of an ellipse equals $C((z^2 - \varepsilon^2)^{1/2} - z)$; it is not rational, so an ellipse is not an algebraic domain.

3. *Answer*: The uniformization map of $D(t)$ is $f_t(\zeta) = a_t\zeta + b_t\zeta^2$ (Figure 37), where a_t and b_t can be expressed in terms of a_0 and b_0 as follows:

$$a_t^2 + 2b_t^2 = a_0^2 + 2b_0^2 - \frac{qt}{\pi}, \qquad a_t^2 b_t = a_0^2 b_0.$$

The time of cusp formation is

$$t^* = \frac{\pi}{q}\left(\frac{3}{2}(2a_0^2 b_0)^{2/3} - a_0^2 - 2b_0^2\right).$$

The quotient of oil that has been extracted by the time t^* equals $\eta = qt^*/\pi(a_0^2 + 2b_0^2)$. The value $\eta = 1/2$ is attained at $b_0/a_0 \approx 0.1$.

4. *Answer*: The uniformization map of the domain $D(t)$ is

$$f_t(\zeta) = \frac{\zeta}{1-\zeta}\left(\frac{4}{3}b + \frac{2}{3}\sqrt{b^2 - 3qt/\pi}\right) - \zeta\left(\frac{1}{3}b - \frac{1}{3}\sqrt{b^2 - 3qt/\pi}\right).$$

At the time of cusp development the velocity $\frac{df}{dt}(-1)$ must be infinite, which happens when the expression under the square root vanishes, i.e. at $t = \pi b^2/3q$.

5. *Answer*: The uniformization map of $D(t)$ has the following form:

$$f_t(\zeta) = a\left(\frac{1}{\pi}\log\frac{1+\zeta}{1-\zeta}\right) + a\zeta\sqrt{\frac{1}{\pi^2} - \frac{qt}{\pi a^2}}.$$

The breakdown time t^* equals $a^2/\pi q$. The amount of fluid that will have been extracted by this time is approximately 2.5 times smaller than in case of extraction from a circle of diameter $a/2$ (Figure 8).

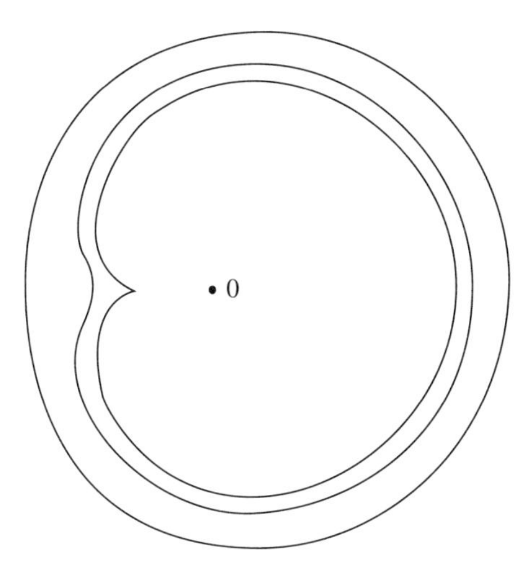

FIGURE 37

6. Assume that the sinks are situated at points $ae^{2\pi ik/N}$, $k = 0, 1, \ldots, N-1$, and let them have an equal rate q. Suppose that the circle is centered at the origin, and let $R > a$ be its radius. Then the uniformization map at a time t equals

$$f_t(\zeta) = \frac{N\beta_t\zeta}{1-\alpha_t^N\zeta^N} + \gamma_t\zeta,$$

where

$$\beta_t = \frac{\frac{\alpha_t}{a}\left(R^2 - \frac{qt}{\pi}\right) + \frac{a}{\alpha_t}}{N+1+(N-1)\alpha_t^{2N}}(1-\alpha_t^{2N}),$$

$$\gamma_t = \frac{(1+(N-1)\alpha_t^{2N})\frac{a}{\alpha_t} + N\frac{\alpha_t}{a}\left(R^2 - \frac{qt}{\pi}\right)}{N+1+(N-1)\alpha_t^{2N}},$$

and α_t can be found from the equation

$$\begin{aligned} R^2(N+1+(N-1)\alpha_t^{2N})^2 \\ = \frac{a^2}{\alpha_t^2}(N+1-\alpha_t^{2N})(1+(N-1)\alpha_t^{2N}) \\ + N\frac{\alpha_t^{2N+2}}{a^2}\left(R^2 - \frac{qt}{\pi}\right)^2 + N\left(R^2 - \frac{qt}{\pi}\right)((N-1)\alpha_t^{4N} + N + 1). \end{aligned}$$

The above formulas make perfect sense if one sets N to be any real number greater than 1. Then one obtains a solution to the problem of extraction of fluid from a sector with angle $2\pi/N$ (Figure 38, p. 64).

7. *Answer*: The domain from Example 1, §3.7.

The moments of the resulting domain do not depend on the radius of the circle along which the source is moving. They depend only on the position of the center, the period of rotation, and the rate of the source. Hence, the

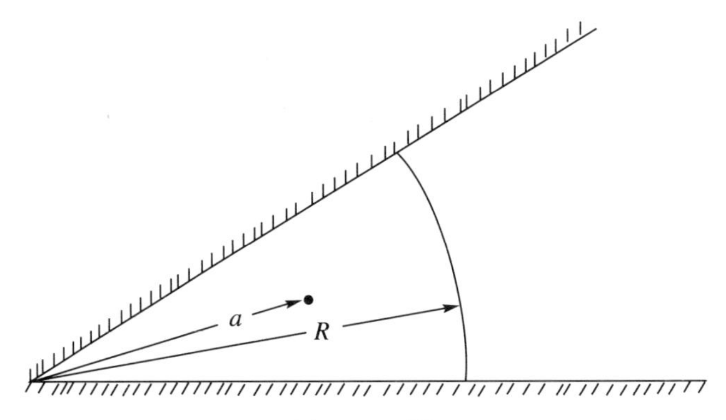

FIGURE 38

result of injection does not depend on the radius of the circle either. But in case of zero radius (i.e. when the source, in fact, is not moving) the domain is the same as in Example 1 in §3.7.

Chapter 4. 1. Since $\bigcup_{t>0} D(t)$ is the entire plane, the intersection $\bigcap_{t>0} \overline{D(t)}$ is empty. Hence, for $k \geq 3$

$$\int_{\overline{D(t)}} z^{-k} dxdy \to 0 \quad \text{as } t \to \infty.$$

By (4.4), this integral is independent of t, which implies

$$\int_{\overline{D(0)}} z^{-k} dxdy = 0, \quad k \geq 3.$$

According to a theorem of M. Sakai (see §4.9), the initial domain has to be an ellipse.

2. Let x_0 be the abscissa of the contraction point. Then for any t

$$\Phi_t(x_1, 0) \geq \Phi_t(x_2, 0) \quad \text{if } x_0 > x_1 > x_2 \text{ or } x_0 < x_1 < x_2$$

(Figure 39, p. 65), since the flow along the horizontal axis is directed towards the contraction point. Hence, if $x_0 > x_1 > x_2$ or $x_0 < x_1 < x_2$ then

$$\int_0^{t^*} \Phi_t(x_1, 0)\, dt > \int_0^{t^*} \Phi_t(x_2, 0)\, dt,$$

or, considering (4.12), $\Pi_{D(0)}(x_1, 0) < \Pi_{D(0)}(x_2, 0)$. Therefore, the contraction point is the only local minimum point of the gravity potential on the abscissa axis.

3. *Hint.* Consider the process of simultaneous contraction of the domain and its images with respect to the rigid boundaries.

Answer: a) The abscissa of the contraction point is $x = \sqrt{b^2 - \rho^2}$, b being the abscissa of the center of the circle, and ρ being its radius (Figure 13).

b) The abscissa of the contraction point is the root of the equation

$$\frac{x-b}{2} + \frac{\rho^2}{2}\frac{\pi}{2d}\left(\cot \pi\frac{x-b}{2d} + \cot \pi\frac{x+b-d}{2d}\right) - \frac{\rho^2}{2}\frac{1}{x-b}$$

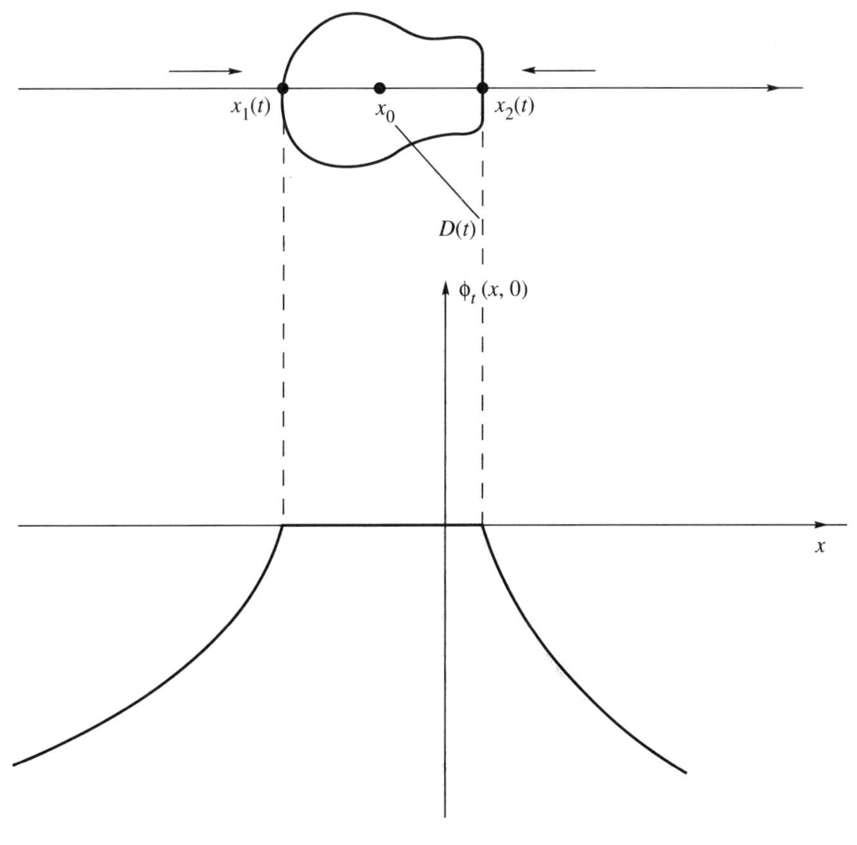

FIGURE 39

that lies between 0 and b. Here ρ is the radius of the circle, d is the width of the channel, and b is the distance from the center of the bubble to the closer wall of the channel (Figure 14).

4. *Answer*: The ratio of half-axes equals $\frac{a}{\sqrt{a^2+4\rho^2}}$.

Chapter 5. 1. The Cauchy transform of a domain bounded by circles is expressed by different rational functions inside different components of the complement. Therefore, the derivative of the Cauchy transform does not extend to a univalent rational function, and the domain is not abelian.

2. The boundary ∂D of an algebraic domain is the image of a real algebraic curve $\partial\Sigma^+$, under a rational mapping, therefore it has to be an algebraic curve. This curve is irreducible.

3. Assume the contrary. Let D be a domain of degree 2, Σ be the corresponding Riemann surface, and let f be the uniformization map of Σ^+ onto D. This map defines a two-fold branched covering of the Riemann sphere by the surface Σ.

Case 1. The surface has genus $g > 1$. Defined on Σ is a hyperelliptic automorphism $\sigma: \Sigma \to \Sigma$ that assigns to an inverse image of a point $z \in \mathbf{C}P^1$

the other inverse image of this point. This automorphism does not depend on f. Let $a \in \mathbf{C}P^1$, $a \notin D$. Then both inverse images λ, μ of a lie in Σ_-; for them, we have $\sigma(\lambda) = \mu$, $\sigma(\mu) = \lambda$, $f(\lambda) = f(\mu)$. Since σ commutes with the antiholomorphic involution τ, we have $\sigma(\tau(\lambda)) = \tau(\mu)$, $\sigma(\tau(\mu)) = \tau(\lambda)$, $f(\tau(\lambda)) = f(\tau(\mu))$. But $\tau(\lambda), \tau(\mu) \in \Sigma^+$. This contradicts the bijectivity of the map f on $\Sigma^+ \cup \partial\Sigma^+$.

Case 2. The surface Σ has genus 1. In this case, the automorphism $\sigma: \Sigma \to \Sigma$ is defined as well, but now it depends upon the mapping f. Let us identify Σ with a torus $\mathbf{C}/\Gamma$, where Γ is a lattice. Then the automorphism will act as follows: $\sigma_f(\zeta) = -\zeta + a(f)$. This automorphism is area-preserving, so $\sigma_f(\Sigma^+ \cup \partial\Sigma^+)$ intersects $\Sigma^+ \cup \partial\Sigma^+$. This contradicts the bijectivity of f on $\Sigma^+ \cup \partial\Sigma^+$.

4. Let D be an algebraic domain, and let

$$h_D(w) = \sum_{j=1}^{n} \sum_{k=1}^{N_j} \frac{a_{jk}}{\pi(w - z_j)^k} = \int_D \frac{dxdy}{w - z}$$

be its Cauchy transform. Since the functions $\frac{1}{w-z}$ considered as functions of z for $w \notin D$ generate linearly the space of all holomorphic functions in D, for an arbitrary holomorphic function ϕ in D we have

$$\int_D \phi(z)\, dx\, dy = \sum_{j=1}^{n} \sum_{k=0}^{N_j - 1} \phi^{(k)}(z_j) \frac{(-1)^k a_{jk}}{k!}.$$

5, 6. Any domain can be approximated by domains with analytic boundary curves. An approximation of a domain with analytic boundaries by algebraic domains can be obtained as follows. First one constructs the Riemann surface Σ that corresponds to the domain (§5.4). Next, one introduces meromorphic functions ϕ_n on Σ, analogues of the functions z^n on the Riemann sphere. They have a pole at a point $P \in \Sigma^-$ and a zero of multiplicity n at $\tau(P) \in \Sigma^-$, and are regular and nonzero elsewhere. Any holomorphic function in Σ^+ expands in a series with respect to this system of functions, which is analogous to the Taylor expansion of a holomorphic function in the circle. This series is uniformly convergent in the closure of Σ^+ if the function is analytic in some neigborhood of the boundary, which holds for a uniformization map of a domain with analytic boundary. A partial sum of this series is a univalent function on Σ^+ provided the number of terms is large enough. It realizes a conformal map of Σ^+ onto a domain close to the initial domain (see **[24]**). The Cauchy transform of this domain is a polynomial of $1/z$.

7. A conformal map of the strip $|\operatorname{Im} z| \leq c$ onto the fluid domain $D(t)$ is given by

$$f_t(w) = A\zeta_\lambda\left(\frac{w}{a} + 2ci\right) - A\zeta_\lambda(2ci) + Bw,$$

where ζ_λ is the Weierstrass function (§5.8), $\lambda = 4c/a$, and the parameters

A, B, and c can be found from the equations

$$A\delta = (1 - B)a; \quad A\delta' = 4i(b - Bc); \quad Aa = -\frac{qt}{\pi m}\frac{1}{\left(B - \frac{A}{a}\zeta_\lambda'\left(\frac{i\lambda}{2}\right)\right)};$$

here δ and δ' are the monodromies of ζ_λ along the periods.

REMARK. This problem is equivalent to the problem of evolution of an annular domain on a cylinder produced by injection from a single source. The machinery developed in Chapter 5 applies in this case as well.

8. During the evolution the potential of the hole changes as follows: $\Pi_{D(t)} = \Pi_{D(0)} + C(t)$, so its gradient is invariant. At the contraction time t^* the domain $D(t)$ is simply connected, and its Cauchy transform equals the Cauchy transform of $D(0)$, i.e. $\frac{R^2}{z} - \frac{\rho^2}{z-a}$. Therefore, $D(t^*)$ is the domain from Example 1 in §3.7. The contraction point can now be found from the identity $\operatorname{grad}\Pi_{D(t^*)}(P) = \operatorname{grad}\Pi_{D(0)}(P)$.

9. Consider the family of domains $D(\varphi)$ obtained from D by rotating it through the angle φ about the origin. The have the same moments. Therefore, they coincide with D.

Chapter 6. 1. If t is large enough, the fluid domain is simply connected, because the inner component of the boundary vanishes at some point. The Cauchy transform of the fluid domain equals

$$h_{D(t)}(w) = \frac{R_2^2 - R_1^2}{w} + \frac{qt}{\pi(w - z_0)}.$$

It is the same as the Cauchy transform of the domain $D(t - \pi R^2)$ from Example 1 in §3.7. It is seen from that example that there exists at most one domain corresponding to a given Cauchy transform with two poles. Therefore, the two domains coincide.

2. *Answer*: $(x^2 + y^2)^2 = P^2x^2 + Q^2y^2$, where

$$P = \sqrt{2\left(\frac{qt}{\pi} + a^2\right)}, \qquad Q = \sqrt{2\left(\frac{qt}{\pi} - a^2\right)}.$$

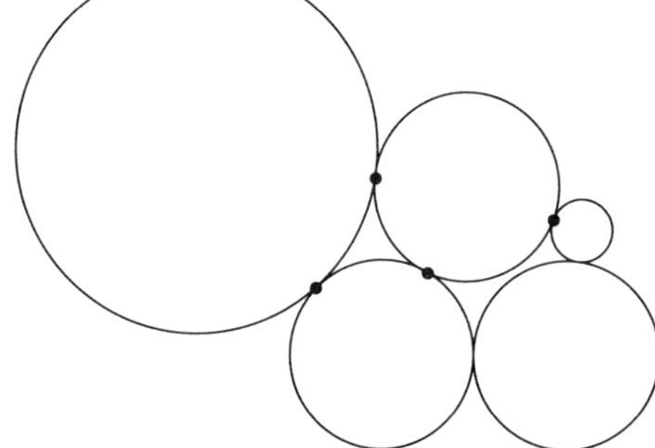

FIGURE 40

3. Choose a connected collection of tangent closed disks with disjoint interiors so that the complement of their union contain $g+1$ connected components (Figure 40, p. 67). The number of disks in such a collection can be made equal to $[(g+1)/2]+2$ since every new disk can be chosen so that it will touch the previous three disks (the circle of Apollonius), which will increase the number of components by two.

Let us now inject small amounts of fluid at the centers z_j of the disks. We will obtain a connected domain with g holes. Its Cauchy transform equals $\sum_{j=1}^{[(g+1)/2]+2} \frac{C_j}{z-z_j}$. Therefore, the domain is algebraic of degree $[(g+1)/2]+2$.

4. *Answer*: $\left(\pm\sqrt{\frac{b(a+b)}{2}-c^2}\,,\,0\right)$.

5. *Answer*: Yes.

Hint: The sufficient condition of a breakup from section 6.4. is valid for domains with piecewise smooth boundaries. The domain shown in Figure 28 breaks up when the ratio of the radii of the outer and inner circles is greater than $e^{\pi/2}$ which approximately equals 4.81.

Chapter 7. 1. *Answer*: Yes.

Hint: The ratio of the surface area to the area of the equatorial section is greater than eight.

2. *Answer*: $\frac{\widehat{\Pi}_{xx}}{\widehat{\Pi}_{yy}}$ at the contraction point $P=(a-\sqrt{a^2+r^2}\,,\,0)$.

Hint: Contraction on a surface as well as contraction on the plane has the property that the axes of the asymptotic ellipse are directed along the eigenvectors of the matrix of second derivatives of the generalized gravity potential at the contraction point, and the lengths of the half-axes are proportional to the corresponding eigenvalues.

A Few Open Questions

Below we give a short list of unsolved questions and conjectures related with the problems we have discussed.

1. Conjecture (P. S. Novikov). *Let D_1 and D_2 be simply connected bounded domains in the plane such that both their intersection and the complement to their union are connected. Then if the outer logarithmic potentials of D_1 and D_2 are the same, the domains are the same as well.*

This conjecture has been proved under various additional restrictions imposed on the domains D_1 and D_2. In 1979 V. Isakov showed that it holds for domains which are convex along a fixed line **[25]**.[1]

2. Question. *Do there exist two distinct simply connected abelian domains with equal outer logarithmic potentials (= equal moments)?*

This question was raised by Aharonov and Shapiro in 1976 **[26]**. C. Ullemar **[27]** showed that such two domains do not exist within the class of domains obtained from the unit disk by a polynomial conformal map of degree 3. There is a hope that the same is true for polynomials of any degree.

Remark. The topological nature of the example of two domains with equal moments might create an impression that such examples must be possible in any dense subclass of simply connected bounded domains. M. Brodsky and V. Strakhov **[28]** showed that it is not so. They proved that the inverse problem of potential theory has a unique solution in the class of domains bounded by level curves of the absolute value of a polynomial of the complex coordinate. Such domains are dense in the set of domains with piecewise smooth boundary.

3. One can formulate an analogue of the evolution problem in the Euclidean space of any dimension. Namely, fix a system of sources $z_1, \dots, z_n$ and rates $q_1, \dots, q_n$, and for any domain D define the velocity potential

$$\Phi(z, D) = \sum_{j=1}^{n} q_j G(z, z_j, D),$$

[1] A domain D is called convex along a line l if the intersection of D with any line parallel to l is either empty or a single interval.

where $G(z, z_j, D)$ is Green's function of D with pole at z_j. Now define the evolution by the condition that the speed of the boundary of D is equal to the normal derivative of $\Phi(z, D)$. Both classical and generalized solutions for this problem can be defined quite similarly to the two-dimensional case. Richardson's theorem has an obvious analogue.

A natural question is:

Can one construct any nontrivial explicit solutions of the evolution problem in the N-dimensional Euclidean space, where $N > 2$? In other words, do there exist any natural multidimensional counterparts of algebraic solutions in the plane?

It is remarkable that the answer is positive if $N = 4$. A family of explicit 4-dimensional solutions was recently discovered by L. Karp [**29**] and independently by the second author of this text. However, for all other dimensions, including the case $N = 3$ which appears in applications, the question still remains open.

4. Theorems 6.4 and 7.4 give sufficient conditions for a symmetric gas region to break up. It is natural to ask under what reasonable assumptions these conditions are also necessary. For instance, will the condition of Theorem 7.4 be necessary if the surface is convex?

5. In contraction theory for nonsymmetric domains, we have found no effectively verifiable sufficient condition for the domain to break up during contraction. The following statement, if it holds, would furnish such a condition.

CONJECTURE. *If the gravity potential of a simply connected domain has more than one local minimum in the plane, then the domain breaks up in the course of evolution.*

Problem 2 in Chapter 4 proves this statement for domains symmetric in the horizontal axis.

It can be shown that it is sufficient to prove the conjecture for domains with exactly two local minima of the potential. To do that, it would be enough just to show that the space of such domains is connected.

References

1. R. Courant, *Dirichlet Principle, Conformal Mapping, and Minimal Surfaces*, Interscience, New York, 1950.
2. Stanley Richardson, *Some Hele-Shaw flows with time-dependent free boundaries*, J. Fluid Mech. **102** (1981), 263–278.
3. P. S. Novikov, *On the uniqueness for the inverse problem of potential theory*, Dokl. Akad. Nauk SSSR **18** (1938), 165–168. (Russian)
4. Makoto Sakai, *A moment problem on Jordan domains*, Proc. Amer. Math. Soc. **70** (1978), 35–38.
5. P. I. Etingof, *Integrability of the problem of filtration with a moving boundary*, Dokl. Akad. Nauk SSSR **313** (1990), 42–47; English transl. in Soviet Phys. Dokl. **35** (1990).
6. B. Gustafsson, *Quadrature identities and the Schottky double*, Acta Appl. Math. **1** (1983), 209–240.
7. Stanley Richardson, *Hele-Shaw flows with a free boundary produced by the injection of fluid into a narrow channel*, J. Fluid Mech. **56** (1972), 609–618.
8. A. I. Markushevich, *Theory of functions of a complex variable,* vol. 1, GITTL, Moscow, 1950; English transl., Prentice Hall, Englewood Cliffs, NJ, 1967 .
9. L. A. Galin, *Unsteady seepage with a free surface*, Dokl. Akad. Nauk SSSR **47** (1945), 250–253. (Russian)
10. B. Gustafsson, *On the differential equation arising in a Hele-Shaw flow moving boundary problem*, Ark. Mat. **22** (1984), 251–268.
11. P. Ya. Polubarinova-Kochina, *On the motion of the oil contour*, Dokl. Akad. Nauk SSSR **47** (1945), 254–257. (Russian)
12. P. P. Kufarev, *The oil contour problem for the circle with any number of wells*, Dokl. Akad. Nauk SSSR **75** (1950), 507–510. (Russian)
13. Makoto Sakai, *Quadrature domains*, Lecture Notes in Math., vol. 934, Springer-Verlag, Berlin and New York, 1982.
14. V. M. Entov and P. I. Etingof, *Bubble contraction in Hele-Shaw cells*, Quart. J. Mech. Appl. Math. **44** (1991), 507–535.
15. Murray H. Protter and Hans F. Weinberger, *Maximum principles in differential equations*, Springer-Verlag, Berlin and New York, 1984.
16. J. W. Milnor, *Singular points of complex hypersurfaces*, Princeton University Press, Princeton, NJ, 1968.
17. Makoto Sakai, *Null quadrature domains*, J. Analyse Math. **40** (1981), 144–154.
18. S. D. Howison, *Bubble growth in porous media and Hele-Shaw cells*, Proc. Royal Soc. of Edinburgh Sect. A **102** (1986), 141–148.
19. A. I. Markushevich, *Theory of functions of a complex variable,* vol. 2, GITTL, Moscow, 1950; English transl., Prentice Hall, Englewood Cliffs, NJ, 1967 .
20. H. Cartan, *Théorie elémentaire des fonctions analytiques d'une ou plusieurs variables complexes*, Hermann, Paris, 1961; English transl., Hermann, Paris, and Addison-Wesley, Reading, MA, 1963 .
21. B. A. Dubrovin, S. P. Novikov, and A. T. Fomenko, *Modern geometry: methods and applications*, 2nd ed., "Nauka", Moscow, 1986; English transl. of 1st ed., Springer-Verlag, Berlin and New York, 1984 .

22. L. Zalcman, *Some inverse problems of potential theory*, Contemp. Math. **63** (1987), 337–350.
23. P. I. Etingof, *Inverse problems of potential theory and flows in porous media with time-dependent free boundary*, Comp. Math. Appl. **22** (1991), 93–99.
24. I. M. Krichever and S. P. Novikov, *Algebras of Virasoro type, Riemann surfaces, and structures of the theory of solitons*, Funktsional. Anal. i Prilozhen. **21** (1987), no. 2, 46–63; English transl., Functional Anal. Appl. **21** (1987), no. 2, 126–142.
25. V. Isakov, *On the uniqueness of a solution in the inverse problem of potential theory*, Dokl. Akad. Nauk SSSR **245** (1979), 1045–1047; English transl. in Soviet Math. Dokl., **20** (1979).
26. D. Aharonov and H. S. Shapiro, *Domains on which analytic functions satisfy quadrature identities*, J. Analyse Math. **30** (1976), 39–73.
27. C. Ullemar, *A uniqueness theorem for domains satisfying quadrature identity for analytic functions*, Preprint TRITA-MAT-1980-37, Mathematics, Royal Inst. of Technology, S-100-44, Stockholm, Sweden.
28. V. N. Strakhov and M. A. Brodsky, *On the uniqueness of the solution of the inverse logarithmic potential problem*, SIAM J. Appl. Math. **46** (1986), 324–344.
29. L. Karp, *Construction of quadrature domains in* $\mathbf{R}^n$ *from quadrature domains in* $\mathbf{R}^2$, Complex Variables: Theory Appl. **17** (1992), 179–188.